Problem Books in Mathematics

Edited by P. R. Halmos

Problem Books in Mathematics

Series Editor: P.R. Halmos

Unsolved Problems in Intuitive Mathematics, Volume I: Unsolved Problems in Number Theory
by *Richard K. Guy*
1981. xviii, 161 pages. 17 illus.

Theorems and Problems in Functional Analysis
by *A.A. Kirillov* and *A.D. Gvishiani* (trans. *Harold H. McFaden*)
1982. ix, 347 pages. 6 illus.

Problems in Analysis
by *Bernard Gelbaum*
1982. vii, 228 pages. 9 illus.

A Problem Seminar
by *Donald J. Newman*
1982. viii, 113 pages.

Donald J. Newman

A Problem Seminar

Springer-Verlag
New York Heidelberg Berlin

Donald J. Newman
Department of Mathematics
Temple University
Philadelphia, PA 19122
U.S.A.

Editor
Paul R. Halmos
Department of Mathematics
Indiana University
Bloomington, IN 47405
U.S.A.

AMS Classification (1980): 00A07

Library of Congress Cataloging in Publication Data
Newman, Donald J., 1930–
A problem seminar.
(Problem books in mathematics)
1. Mathematics—Problems, exercises, etc. 2. Problem
solving. I. Title.
QA43.N43 1982 510′.76 82-16729

Typeset by Science Typographers, Inc., Medford, N Y.

9 8 7 6 5 4 3 2 1

ISBN-13:978-0-387-90765-9 e-ISBN-13:978-1-4613-8214-0
DOI: 10.1007/978-1-4613-8214-0

PREFACE

There was once a bumper sticker that read, "Remember the good old days when *air* was clean and *sex* was dirty?" Indeed, some of us are old enough to remember not only *those* good old days, but even the days when Math was *fun*(!), not the ponderous THEOREM, PROOF, THEOREM, PROOF,..., but the whimsical, "I've got a good problem."

Why did the mood change? What misguided educational philosophy transformed graduate mathematics from a passionate activity to a form of passive scholarship?

In less sentimental terms, why have the graduate schools dropped the Problem Seminar? We therefore offer "A Problem Seminar" to those students who haven't enjoyed the fun and games of problem solving.

CONTENTS

FORMAT

This book has three parts: first, the list of problems, briefly punctuated by some descriptive pages; second, a list of hints, which are merely meant as words to the (very) wise; and third, the (almost) complete solutions. Thus, the problems can be viewed on any of three levels: as somewhat difficult challenges (without the hints), as more routine problems (with the hints), or as a textbook on "how to solve it" (when the solutions are read). Of course it is our hope that the book can be enjoyed on any of these three levels.

PROBLEMS

1. Derive the operations $+$, $-$, $\times$, and $\div$ from $-$ and reciprocal.

2. Invent a single (binary) operation from which $+$, $-$, $\times$, and $\div$ can be derived.

3. The multiplication of two complex numbers $(a + bi)\cdot(x + yi) = ax - by + (bx + ay)i$ appears to need 4 real multiplications $(a \cdot x, b \cdot y, b \cdot x, a \cdot y)$, but does it really? If additions are free, can this same job be accomplished in 3 real multiplications? In 2?

4. A microbe either splits into two perfect copies of itself or else disintegrates. If the probability of splitting is p, what is the probability that one microbe will produce an everlasting colony?

5. Given any n distinct points in the plane, show that one of the angles determined by them is $\leqslant \pi/n$ (the 0 angle counts).

6. Prove that every sequence (of real numbers) contains a monotone subsequence.

7. Suppose $f(x) = x + x^2/n$, and form $\underbrace{f(f(f\ldots f(x))\ldots)}_{n}$ (same n, n a positive integer). What is the limiting behavior as $n \to \infty$?

8. Devise an experiment which uses only tosses of a fair coin, but which has success probability $\frac{1}{3}$. Do the same for any success probability p, $0 \leqslant p \leqslant 1$.

9. We alternate writing down binary digits after a decimal point, thereby producing a real number in the interval $[0, 1]$. You win if this number is transcendental. Can you force a win?

10. At a certain corner, the traffic light is green for 30 seconds and then red for 30 seconds. On the average, how much time is lost at this corner?

11. Prove that there is no equilateral triangle all of whose vertices are plane lattice points. (How about three-dimensional lattice points?)

12. Prove that a sequence of positive numbers, each of which is less than the average of the previous two, is convergent.

13. $x_{n+1} = \frac{1}{2}(x_n + 1/x_n)$, x_0 a given *complex* number. Discuss convergence.

14. $x_{n+1} = (x_n + x_{n-1})/2$, x_0, x_1 given. Express $\lim x_n$ explicitly.

15. If a set of positive integers has sum n, what is the biggest its product can be?

16. Given a convergent series of positive terms, Σa_n, prove that $\Sigma\sqrt[n]{a_1 a_2 \ldots a_n}$ must also be convergent.

17. What is the lowest degree monic polynomial which vanishes identically on the integers (mod 100)? (And generally (mod n)?)

18. Evaluate $\sqrt{1+2\sqrt{1+3\sqrt{1+4\sqrt{1+\cdots}}}}$.

19. Prove that, at any party, two people have the same number of friends present.

20. If n is any integer greater than 1, then n does *not* divide $2^n - 1$.

21. Prove that every non-multiple of 3 is congruent to a power of 2 (mod 3^n).

22. How many perfect squares are there (mod 2^n)?

23. Maximize $2^{-x} + 2^{-1/x}$ over $(0, \infty)$.

24. N distinct non-collinear points are given. Prove that they determine at least N distinct lines.

25. Given that $f(x)$ increases from 0 to 1 as x does, prove that the graph of $y = f(x)$ $(0 \leqslant x \leqslant 1)$ can be covered by n rectangles with sides parallel to the axes and each having area $1/n^2$.

26. Given a finite collection of closed squares of total area 3, prove that they can be arranged to cover the unit square.

27. Given a finite collection of squares of total area $\frac{1}{2}$, show that they can be arranged so as to fit in a unit square (with no overlaps).

28. Devise the smallest plane set such that no point is at a rational distance from all points of the set.

29. Given an infinite number of points in the plane with all the mutual distances integers, prove that the points are all collinear.

30. If a, b are positive integers, then $(a+\frac{1}{2})^n + (b+\frac{1}{2})^n$ is an integer only for finitely many positive integers, n.

31. Given that $f(x, y)$ is a polynomial in x for each fixed y, and $f(x, y)$ is a polynomial in y for each fixed x, must $f(x, y)$ be a polynomial in x and y?

32. Prove that the product of 3 consecutive integers is never a perfect power (i.e., a perfect square or a perfect cube, etc.).

33. Given a region whose boundary is a simple polygon, prove that it contains a disc with radius larger than area/perimeter.

34. I choose an integer from 0 through 15. You ask me 7 yes or no questions. I answer them all, but I am allowed to *lie* once. (I needn't, but I am allowed to.) Determine my number!

35. Given any bounded plane region, prove that there are three *concurrent* lines that cut it into six pieces of equal area.

36. Given any bounded plane region, prove that there is a point through which no line trisects the area.

37. $a, b, c, d, \ldots$ are positive numbers. Prove

$$\sqrt{a+b+c+d+\cdots}+\sqrt{b+c+d+\cdots}+\sqrt{c+d+\cdots}+\cdots$$
$$\geqslant \sqrt{a+4b+9c+16d+\cdots}\,.$$

38. Show that the number 16 is a perfect *8th* power (mod p) for any prime p.

39. The points of the plane are each colored either red, yellow, or blue. Prove that there are two points of the same color having mutual distance 1.

40. Assume that the points of the plane are each colored red or blue. Prove that one of these colors contains pairs of points at *every* mutual distance.

41. Given a simple plane arc of length more than 1, prove that for some n there are more than n points on the arc whose mutual distances are all at least $1/n$.

42. Good coins weigh 10 gm, bad ones 9 gm. Given 4 coins and a scale (not a balance, but a true scale), determine which are which in only 3 weighings.

43. Let $[\alpha, \beta]$ be an interval which contains no integers. Show that there is a positive integer n such that $[n\alpha, n\beta]$ still contains no integers but has length at least $\frac{1}{6}$.

44. Prove that the integers $[(\sqrt{2}+1)^n]$ are alternately even and odd.

45. Prove that $\min_k(k+[n/k]) = [\sqrt{4n+1}\,]$. (Here n is a given positive integer and k varies over all positive integers.)

46. Let α, β be positive irrationals. Show that the sets $[n\alpha]$ and $[n\beta]$, $n = 1, 2, 3, \ldots$, are complements iff $1/\alpha + 1/\beta = 1$.

47. Suppose we "sieve" the integers as follows: we choose $a_1 = 1$ and then delete $a_1 + 1 = 2$. The next term is 3, which we call a_2, and then we delete $a_2 + 2 = 5$. Thus, the next available integer is $4 = a_3$, and we delete $a_3 + 3 = 7$, etc. Thereby we leave the integers $1, 3, 4, 6, 8, 9, 11, 12, 14, 16, 17, \ldots$. Find a formula for a_n.

48. Call an integer square-full if each of its prime factors occurs to the second power (at least). Prove that there are infinitely many pairs of consecutive square-fulls.

49. We term two sets "almost disjoint" if their intersection is finite. What is the largest (cardinality) collection of sets of integers which are pairwise almost disjoint?

50. Split a beer *three ways*. To split a beer *two ways* you let the first man divide it into what he thinks are two equal parts and then let the second man choose one of them. Both are then satisfied. How can three do this?

51. Define x_n by $x_n = x_{n-1} + \frac{1}{2}x_{n-2}$, $x_0 = 0$, $x_1 = 1$. Prove that for $n > 8$, x_n is not an integer.

52. Let $a_1, a_2, \ldots, a_k$ be positive integers and let S be the set of all positive integers not divisible by any of them. Prove that the density of S is at least $(1-1/a_1)(1-1/a_2)\ldots(1-1/a_k)$.

53. Given any n real pairs (x_i, y_i), with the x_i all distinct, prove that the interpolation problem $P(x_i) = y_i$, $i = 1, 2, \ldots, n$, can be solved by a polynomial P, all of whose zeros are real.

54. Is there a non-trivial function $f(x)$, continuous on the whole line, which satisfies the functional equation $f(x) + f(2x) + f(3x) \equiv 0$?

55. Give examples of:
(1) An infinite group with no infinite *proper* subgroup.
(2) A field isomorphic to a proper subfield.
(3) A ring with no maximal ideals.

56. $n_1, n_2, n_3, \ldots$ is a sequence of positive integers with the property $n_{k+1} > n_{n_k}$. Show that it must be the sequence $1, 2, 3, 4, \ldots$.

57. (A CHEERFUL FACT ABOUT...) Given a right triangle and a finite set of points inside it, prove that these points can be

connected by a path of line segments the sum of whose squares is bounded by *the square of the hypotenuse*.

58. Batter A has a higher batting average than batter B for the first half of the season *and* A also has a higher batting average than B for the second half of the season. Does it follow that A has a better batting average than B for the whole season?

Estimation Theory

This whole topic is perhaps born out of the shortage of exact formulas. In many (most?) situations a quantity is sought which cannot be expressed in simple closed form, but can only be estimated. Of course the game is to estimate it *well*, but how? That is, how do you know that you've done a good job?

The answer is that this game is played on two "fronts." One estimates above (finds something definitely larger than the desired quantity), and one also estimates below (finds something definitely smaller). In many cases the arguments leading to these two bounds may be totally unrelated and quite ad hoc. Each argument *alone* gives no assurance that it is getting anywhere near the truth—*BUT*, when the two arguments give results which are close to one another, they PROVE each other out. It is then that one knows he has been *clever* and accurate!

There is a nice freedom in this subject. Basically there are no *wrong* answers, just better or worse ones depending on how close the resultant upper and lower bounds turn out to be.

59. If $e^{e^x} = \Sigma a_n x^n$, estimate a_n.

60. If $x_0 = 1$, $x_{n+1} = x_n + 1/x_n$, then $x_n \to \infty$ (why?). Estimate how fast.

61. $\Sigma_{n=0}^{\infty} x^{n^2}$ tends to ∞ as $x \to 1^-$. How fast?

62. The equation $x^n + x = 1$ has a unique positive solution $x(n)$, which approaches 1 as $n \to \infty$. Estimate how fast.

63. $\underbrace{\sin \sin \sin \ldots \sin}_{n}(\pi/2)$ goes to 0 as $n \to \infty$. Estimate how fast.

64. The function $f(x) = \Sigma_{n=0}^{\infty} 1/(2^n + x)$ goes to 0 as $x \to \infty$. How fast?

65. Estimate the largest collection of triples one can choose from n elements such that no two of them overlap in more than one element.

66. Consider the sequence $1, 2, 3, 4, 6, 8, 9, 12, 16, 18, \ldots$ of positive integers composed of 2's and 3's (that is of the form $2^a 3^b$, a, b non-negative integers) and arranged in increasing order. Prove that the ratio of successive terms approaches 1.

67. Find, asymptotically, how many lattice points in the square $0 < x \leqslant N$, $0 < y \leqslant N$, are "visible" from the origin. (The point $(5,8)$ is visible from the origin since no other lattice point blocks the view, whereas the point $(6,8)$ is blocked by the point $(3,4)$, i.e., the line from the origin to $(6,8)$ hits $(3,4)$.)

68. Find, asymptotically, the number of lattice points in the disc $x^2 + y^2 \leqslant R^2$ as $R \to \infty$.

69. Suppose we are given a triangle whose vertices are lattice points, but which contains no other lattice points in or on it. Using asymptotics, tiling the plane with copies of the given triangle, deduce that the area of the triangle is $\frac{1}{2}$.

70. The series $\Sigma_{n \in S}(x^n/n!)$ is e^x when S is all of the (non-negative) integers, is $\cosh x \sim \frac{1}{2}e^x$ when S is the even integers, and is $\sinh x \sim \frac{1}{2}e^x$ when S is the odd integers. What is it asymptotically when S is something else? For example, the multiples of 3, of 4, of 5,...,etc.

71. Estimate the size of the largest subset of $\{1,2,3,\ldots,n\}$ which doesn't contain any term equal to double another term.

72. Consider the sequence $\{X_n\}$ comprising all perfect powers (squares, cubes, etc.) arranged in increasing order (so it begins $1,4,8,9,16,25,27,\ldots$). Find X_n asymptotically.

73. Given n points in the unit square, there is a shortest curve connecting them. Estimate the longest this curve can be.

74. For a and b positive integers less than n, estimate the *longest* that the Euclidean algorithm can take (to determine $\gcd(a, b)$).

75. Given n points, what is the maximum number of connecting lines that can be drawn without producing a triangle? (Note that

the only triangles we count are those with all three vertices among the given points.)

76. Prove $n-[n/2]+[n/3]-\cdots \sim n \log 2$.

77. Let $P(x, y)$ be a polynomial of x degree m, and y degree n. Prove that $P(x, e^x)$ can have at most $mn+m+n$ real zeros.

78. Show that $f(x) \sim x^2$ (as $x \to \infty$) does not imply $f'(x) \sim 2x$, but that it *does* for convex $f(x)$.

79. *A New l'Hôpital's Rule*: Suppose $f(x)$, $g(x)$ are differentiable on $(0,1]$, that $g'(x)>0$ there, and that $\lim_{x \to 0^+}(f'(x)/g'(x))$ exists (on the extended reals). Then, *whether or not* it is an indeterminate form, $\lim_{x \to 0^+}(f(x)/g(x))$ exists (again on the extended reals).

80. Given that $f(x)+f'(x) \to 0$ as $x \to \infty$, prove that both $f(x) \to 0$ and $f'(x) \to 0$.

81. $F(x)$ is a positive increasing function on $[0, \infty)$ and y is any solution to the differential equation $y''+F(x)y=0$. Prove y remains bounded as $x \to \infty$.

82. Show that if $f(x)$ and $f''(x)$ are bounded, then $f'(x)$ is. (Here $f(x) \in C^2$, and the domain is the whole line.)

83. Prove that the equation $x^{x^{x^{\cdot^{\cdot^{\cdot}}}}}=2$ is satisfied by $x=\sqrt{2}$, but that the equation $x^{x^{x^{\cdot^{\cdot^{\cdot}}}}}=4$ has no solution. What is the "break-point"?

84. $f(x)$ is continuous and satisfies

$$\lim_{h \to 0^+} [(f(x+2h) - f(x+h))/h] = 0$$

for each x. Prove that $f(x)$ is a constant.

85. Show that $f(x) \in C^1[a, b]$ iff the limit as $h \to 0$ of

$$(f(x+h) - f(x))/h$$

exists *uniformly* on $[a, b]$.

86. Let $f(\theta) = \sin\theta \sin 2\theta \sin 4\theta \ldots \sin 2^n\theta$. Prove that $|f(\theta)| \leqslant (2/\sqrt{3})|f(\pi/3)|$. (The function is "nearly" maximized at $\pi/3$.)

87. $\{A_n\}$ is a sequence of positive numbers satisfying $A_n < A_{n+1} + A_{n^2}$ for all n. Prove that $\sum A_n$ diverges.

Generating Functions

This is one of the "must-see's" of mathematics: Euler's magnificent idea of examining the integers by looking at power series. The gist of the idea is that adding the integers a and b **corresponds** to multiplying the powers x^a and x^b. By exploiting this simple correspondence one can translate many problems in number theory into problems about power series, and (sometimes) an answer can thereby be found.

A good example is that of the binary expansion. The number theoretic statement is that every positive integer has exactly one representation as the sum of distinct powers of 2. The corresponding power series formulation is obtained by looking at $(1+x)(1+x^2)(1+x^4)(1+x^8)\ldots$, which multiplies out to give terms like $x^1 \cdot x^4 \cdot x^8 = x^{1+4+8}$, in short all the terms x^K where K is the sum of distinct powers of 2. Thus the translated problem reads as the identity $(1+x)(1+x^2)(1+x^4)(1+x^8)\ldots = 1 + x + x^2 + x^3 + \cdots$.

The point is that these translated problems can then be treated by function theoretic methods. Thus, the above identity, which can be written $(1+x)(1+x^2)(1+x^4)\ldots = 1/(1-x)$ or $(1-x)(1+x)(1+x^2)\cdots = 1$, is established easily when it is observed that $(1-x)(1+x)(1+x^2)(1+x^4)\cdots = (1-x^2)(1+x^2)(1+x^4)\ldots$ (i.e., $f(x) = (1-x)(1+x)(1+x^2)\ldots$ satisfies $f(x) = f(x^2)$ so that $f(x)$ is a constant, etc.).

88. *CRAZY DICE*. Devise a pair of dice, cubes with positive integers on their faces, with exactly the same outcomes as ordinary dice (the sum 2 comes out once, the sum 3 comes out twice, etc.), but which are not ordinary dice.

89. Partition the non-negative integers into two sets, A and B, such that every **positive** integer is expressible by $a + a'$; $a < a'$; $a, a' \in A$ in the same number of ways as by $b + b'$; $b < b'$; $b, b' \in B$.

90. Can the positive integers be partitioned into at least two arithmetic progressions such that they all have different common differences?

91. Does there exist an infinite set of positive integers such that all *large* integers are expressible as the sum of two of them in the same number of ways? (Order does not count.)

92. The numbers $0, 2, 5, 6$ have the property that their positive differences are the numbers $1, 2, 3, 4, 5, 6$ each taken on once. Can this phenomenon occur for some number above 6?

93. Prove that the number of partitions of an integer into **odd** positive integers is equal to the number of partitions into distinct positive integers.

Limits of Integrals

An eternal task in analytical problems is that of passing to the limit under the integral sign. The lesson we have all learned is that justification is usually best given by dominated convergence. That is, $\lim_n \int f_n = \int \lim f_n$ provided $\int \operatorname{Sup}_n |f_n| < \infty$. A useful point of view in this regard is that if f_n has some sort of "formula" in terms of n then this Sup can be obtained by ordinary calculus techniques, and the requisite justification can thereby be obtained.

So, although it is not usually thought of in this way, we will view the evaluation of $F(x) = \operatorname{Sup}|f_n(x)|$ as a calculus problem and then simply test whether $\int F(x) < \infty$.

94. Estimate (asymptotically?) $\int_0^\infty (1+x/n)^n e^{-x}\,dx$ as $n \to \infty$.

95. Stirling's Formula: Show that $\int_0^\infty x^n e^{-x}\,dx \sim (n/e)^n \sqrt{2\pi n}$.

96. Show that $1 + n/1! + n^2/2! + \cdots n^n/n! \sim \frac{1}{2}e^n$.

97. Estimate $\int_0^1 \cos^n x^2\,dx$ as $n \to \infty$.

Expectations

A central notion in probability theory is that of the *expectation* or average value. Picture an experiment which consists of several stages, and which continues until a certain outcome is obtained. We wish to determine, on the average, how many stages (trials) are necessary for this to result. (For example if we toss a coin repeatedly until a HEAD first shows up then, although this can happen in 1 toss or take as many as 10,000, the average can be shown to be 2.)

The result which will prove very handy to us in this regard is that this expectation, or average number, is simply equal to $f_0 + f_1 + f_2 + \cdots f_n + \cdots$, where f_n is the probability that the experiment fails to produce the desired outcome for n steps. Thus, $f_0 = 1$ in all cases and in the previous example $f_n \equiv 1/2^n$ so that the sum there is indeed 2. The proof is usually given by a rearrangement of the defining series for the expectation. A more direct argument, however, results from the fact that expectations are "additive." Thus, since the expected amount of time spent on the nth trial is equal to the probability that the nth trial occurs, and since this is equal to f_{n-1}, we do obtain the net expectation of $f_0 + f_1 + f_2 + \cdots$ as asserted.

98. We play the coin tossing game. (If our tosses match, I get both coins; if they differ, you get both.) You have m coins, I have n. What is the expected length of the game (number of tosses until one of us is wiped out)?

99. Suppose you are playing the usual coin tossing game for \$1 per toss, but that the coin is loaded 51% to 49% in your favor. What is the maximum amount you can *expect* to be behind?

100. A "continuous" roulette wheel has all numbers from 0 to 1. We repeatedly spin this until the numbers that arise add up to at least 1. What is the expected number of spins?

101. There are n equally likely alternatives (birthdays). Independent samples are chosen until a repeat occurs. What is the expected number of these choices? What are the asymptotics?

102. A gardener plants n flowers. Each flower takes root with probability $\frac{1}{2}$. The next day those which did not take root are replanted. The process continues until all n take root. What is the expected number of planting days? Asymptotically?

103. Show that any coin toss experiment with a success probability of $\frac{1}{3}$ has an expected number of tosses of at least 2.

104. n letters are placed at random into n envelopes. What is the expected number of letters which get into the correct envelopes?

Prime Factors

The following problem would be far beyond our abilities without a bit of knowledge regarding the "general behavior" of prime factorization. We give the statement in heuristic language although the e, δ form can certainly be framed by the reader, FACT: "Most" numbers n have the number of prime factors "asymptotic" to $\log\log n$ (and this is true whether or not we count the multiplicity of these factors).

We choose to count the multiplicity. So, for example, 72 would have 5 prime factors (far beyond $\log\log 72 \simeq 1.5$). A number up around a googol (10^{100}) usually has around 5 prime factors, and one near a googol-plex ($10^{10^{100}}$) has around 232.

The proof of this fact is not really difficult, but most students would probably not have seen it, even in a course in number theory. At any rate we take it as correct, and turn to our problem.

105. On the multiplication table of numbers 1 through n times numbers 1 through n, show that "almost none" of the numbers (1 through n^2) actually appear!

Category Arguments

There are a bunch of results in real variables which are on a somewhat less trivial level than most. These are the ones which are given by the so-called "category argument". The simplest form of this argument notes that the nested interval theorem holds even for open intervals provided that they do not share any endpoints. The reason being that in that case these open intervals can each be shrunk to smaller non-trivial closed intervals which are still nested. Another one of this circle of results is that a sequence of *dense open* sets has a non-empty intersection. We see this since we can easily produce the sequence of "properly" nested open intervals, one from each of these dense open sets. Finally, by looking at the complements of these *open* sets, we obtain a result for *closed* sets. Namely, if a sequence of closed sets has the whole line for its union then one of the sets already contains a non-trivial interval.

106. A function $f(x) \in C[0, \infty]$ is called *slow* if $f(x+a)-f(x) \to 0$ as $x \to \infty$ for each fixed a. Prove that a slow function can be written as a sum $g(x)+h(x)$, where $g(x) \to 0$ and $h'(x) \to 0$ as $x \to \infty$.

107. $f(x)$ is continuous on $[0, \infty)$, and is such that, for each fixed $a > 0$, $f(na) \to 0$. Must $f(x) \to 0$ as $x \to \infty$?

Convexity

The idea of convex sets is one with rather extensive generality. Although we tend to think of a convex set as a picture we can draw on a blackboard, the notion is not at all limited to two dimensions, or even finitely many dimensions. The general backdrop is a linear space with a topology attached, and there is then the very important concept of *extreme point*, which means a point which is not a convex combination of other points of the set.

The basic theorem (which will prove useful to us) is that a compact convex set always contains extreme points. Enough of them, in fact, so that the whole set is the closure of their convex combinations.

Actually we will be involved with what are called convex *cones* rather than convex sets. These are defined as being closed under taking positive linear combinations rather than only such combinations with coefficients summing to 1. The concepts of extreme points, etc., go over lock, stock, and barrel.

108. A function $f(x) \in C^{\infty}[0, \infty)$ is called *completely monotonic* if $(-d/dx)^k f(x) \geqslant 0$ for all $k = 0, 1, 2, \ldots$ and all $x \geqslant 0$. These functions form a convex cone. Show that the extreme points are the functions $\alpha e^{-\beta x}$, $\alpha, \beta \geqslant 0$.

109. At each plane lattice point there is placed a positive number in such a way that each is the average of its four nearest neighbors. Show that all the numbers are the same!

HINTS

1. Try to derive the operation of *squaring*. (This will then yield products, and so also quotients.)

2. The previous result tells us that we need only generate $1/$ and $-$, so try an operation which *involves* these.

3. Try $(a + b)(x + y)$ in conjunction with two other products.

4. Express this unknown probability in terms of itself by noting that success can only occur if a split occurs at first and then if at least one daughter succeeds.

5. Think of the n points forming a convex k-gon with $n - k$ points inside.

6. Either there are infinitely many "large" terms or there are not.

7. Note that the recurrence relation $y_{k+1} = y_k + (1/n)y_k^2$ is reminiscent of the finite difference method for the differential equation $y' = y^2$, etc.

8. Write the binary expansion of p and then think of these separate powers of $\frac{1}{2}$ as probabilities.

9. Enumerate the algebraics and block each one of them with one of your moves.

10. Write a recurrence formula for this expectation and observe that it has a quadratic solution, etc.

11. Evaluate such a triangle's area two different ways: by the determinant and equilateral formulas.

12. Look at the new sequence $y_n = \mathrm{Max}(x_n, x_{n+1})$, x_n the given sequence, and observe that it is monotone.

13. Exploit the identity

$$\frac{1+z^2}{1-z^2} = \frac{1}{2}\left(\frac{1+z}{1-z} + \frac{1-z}{1+z}\right)$$

to obtain the explicit solution.

14. Write the general solution and pass to the limit.

15. The continuous analogue would call for the numbers to be (nearly) e. Try 3's instead.

16. Use the arithmetic-geometric inequality on $n \cdot a_n$.

17. Write all polynomials in the form $c_0 + c_1\binom{x}{1} + c_2\binom{x}{2} + \cdots$, etc.

18. Consider $f(x) = \sqrt{1 + x\sqrt{1 + (x+1)\sqrt{1 + \cdots}}}$, show that $f^2(x) = 1 + xf(x+1)$, and use estimation methods to deduce that $f(x) = x + 1$.

20. Use the fact that if p is a prime and $(n, p-1) = 1$ then $p \nmid (2^n - 1)$.

21. Use induction to prove that $2^x \equiv 1(3^n)$ does not occur until $x = 2 \cdot 3^{n-1}$.

22. First determine the number of odd squares in this system by noting that each such becomes two on going from n to $n+1$, $n \geqslant 3$.

23. At the maximum point the derivative condition gives $2^{-1/x} = x^2 2^{-x}$ so we can replace $2^{-1/x}$ by this simpler expression and need only prove a simpler inequality.

24. Employ linear algebra! Assign non-trivial numbers to the points so that the total on each line is 0, etc.

25. The first $n-1$ are simply chosen in turn by the requirement of having area $1/n^2$. Then use the Schwarz inequality to show that the remaining rectangle has area $\leqslant 1/n^2$.

26. Arrange them in decreasing order, line them up in rows of lengths just above 1, and pile these rows one on another so as to leave no spaces.

27. Arrange them in decreasing order, line them up in rows of length just below 1, and pile these rows one on another so as to have no overlaps.

28. Try the three points $(0,0)$, $(1,0)$, and $(\xi,0)$. (You pick ξ.)

29. If A, B are any two of these points then all the others lie on a finite number of hyperbolas with axes along $\overline{\mathrm{AB}}$. Ditto for B, C (A, B, C not collinear), etc.

31. For infinitely many y, the polynomial (in x) $f(x, y)$ must have a bounded degree. But then the coefficients would be polynomials in y, and this would force the whole thing to be a polynomial.

32. If $(n-1)n(n+1)$ is a kth power then n^2-1 and n^2 are both kth powers; impossible!

33. If there were no disc of radius r then the strips of width r along the sides would exhaust the area, etc.

34. We must determine 4 binary digits. Ask for the first 3 and then, for the 4th question, ask "Have you lied yet?," etc.

36. Just use the result of Problem 35.

37. Write $a = A^2$, $b = B^2$, etc., and interpret the terms as Euclidean distances.

38. If 2 and -2 are both non-residues, then -4 is a residue, etc.

41. First do the problem for a polygonal arc by removing neighborhoods of the vertices and picking an n so large that the resulting line segments are all at least $1/n$ apart from each other, etc.

44. $(\sqrt{2}+1)^n + (-\sqrt{2}+1)^n = 2\Sigma\binom{n}{2k}2^k$ is an *even* integer.

45. The function $[x + n/x]$ is unimodal, so the integer that minimizes is one of the two integers flanking the real number which minimizes.

46. Note that $1/\alpha$ and $1/\beta$ are the densities of these two sets. This gives the necessity, but a careful count of their numbers gives the sufficiency, too.

47. Problem 46 tells us that $\{[\alpha n + n]\}$ is complementary to $\{[\alpha n]\}$ if $1/\alpha + 1/(\alpha+1) = 1$ with α irrational, i.e., if $\alpha = (\sqrt{5}+1)/2$. Conclude, then, that $a_n = [((\sqrt{5}+1)/2)n]$.

49. Think of sequences of rationals instead of sets of integers! We can take such a sequence converging to any real number.

50. Let the first man pour what he would be satisfied with, but let the other two then take one turn each to rectify it, etc.

51. Try using the exact formula for the progressions $4n, 4n+1, 4n+2, 4n+3$.

52. Try induction, noting that the "sieved set," S (set remaining after all multiples of $a_1, a_2, \ldots$ have been removed), is closed under passing to a divisor, i.e., $n \in S, d|n \Rightarrow d \in S$.

53. Insert more data so that the y_i alternate in sign.

54. Try $f(x) = |x|^\alpha$, α some complex number (with $\operatorname{Re}\alpha > 0$ to ensure continuity).

57. Cut the triangle in two by dropping the altitude and then use induction.

59. Use $a_n x^n \leqslant e^{e^x}$ to get an upper bound and then use (roughly) $a_n \geqslant$ the x^n coefficient of $e^{x^k/k!}$ to get a lower one.

60. Square both sides to get $y_{n+1} = y_n + 2 + 1/y_n$, where $y_n = x_n^2$.

61. Compare the sum with a definite integral.

62. Observe that if $a^n + a > 1$, then a serves as an upper bound, while if $b^n + b < 1$ then b serves as a lower bound.

63. This is just the recurrence sequence defined by $x_{n+1} = \sin x_n$. Try to imitate Problem 60.

64. Compare the sum with a definite integral.

66. This ratio is $\leqslant 2$ from the beginning, and is $\leqslant \frac{3}{2}$ from the second term on. Generalize.

67. If we call their number $f(N)$ then note that $f(N)+f(N/2)+f(N/3)+\cdots$ counts *all* the lattice points in the square. Solve for $f(N)$, etc.

68. Each lattice point accounts for 1 unit of area, so the number of such is nearly equal to the total area.

69. Tile the plane, then notice that the number of triangles is asymptotically twice the number of lattice points, and so twice the area.

70. Use the expansions of $e^{\omega x}$, ω a root of unity, to obtain closed form expressions for these.

71. Think first of the odds, then toss in the odd multiples of 4, then of 16, etc.

72. Try instead to find the "counting function" (number of x_n below x) asymptotically.

73. The lattice gridwork requires a length of around $\sqrt{n}$ (so this is a lower bound). On the other hand, by hitting these "lattice" points and making slight detours, we can connect any n points (so this gives the upper bound).

75. The restriction means that if two points are connected then they cannot both be connected to any other point.

76. Observe that the terms with denominators from $n/2$ to n are all ± 1, those from $n/3$ to $n/2$ are all ± 2, etc.

77. Use Rolle's theorem. Thus, take the $(m+1)$st derivative, divide out e^x and thereby reduce n to $n-1$. Induction gives $mn+n-1$ zeros, and Rolle allows only $m+1$ more.

78. Form the difference quotient, $(f(x+h)-f(x))/h$, but do not let $h \to 0$. Use the convex function inequalities instead.

79. Observe that the limit need only exist through the sequence of *critical* points (where the derivative is 0) for it to exist fully.

80. Use L'Hôpital's rule on $f(x)e^x/e^x$.

81. Integrate the expression $2y'y''/F(x)+2y'y$ $(=0)$ by the *third* mean value theorem.

82. Consider the graph $y=f'(x)$, so that $f(x)$ is the running area and $f''(x)$ is the slope.

83. As a recurrent sequence we have $x_{n+1}=\sqrt{2}^{\,x_n}$, and monotonicity gives convergence, etc.

84. If $f(a)\neq f(b)$, then a linear function could be subtracted off to restore $f(a)=f(b)$, and then the maximum point (in $[a,b)$) would lead to a contradiction.

86. Try forcing $g(\theta)=|\sin\theta|\,|\sin 2\theta|^{\alpha}$ to take its maximum at $\pi/3$ by a shrewd choice of α, and then reconstruct $|f(\theta)|$ from $g(\theta), g(2\theta), g(4\theta),\ldots$.

87. Use the hypothesis repeatedly to obtain $a_2 < a_3 + a_4 < a_4 + a_5 + a_9 + a_{16} < a_{5,6,10,16,17,25,81,256,\ldots}$. Show that a *repeat* never occurs and deduce that the "tails" of the series stay above a_2.

88. The generating function form of the problem asks for a factorization of $(x+x^2+\cdots x^6)^2$ other than the obvious one. So observe that $x+x^2+\cdots x^6$ is $x(1+x)(1+x+x^2)(1-x+x^2)$.

89. The generating function form of the problem is

$$\frac{A^2(x)-A(x^2)}{2}=\frac{B^2(x)-B(x^2)}{2},$$

where $A(x)=\Sigma_{a\in A}x^a$, $B(x)=\Sigma_{b\in B}x^b$. Since $A(x)+B(x)=1/(1-x)$, this gives $A(x)-B(x)=(1-x)(A(x^2)-B(x^2))$. Iterate this relation, etc.

90. The generating function form of the problem is

$$\frac{x}{1-x}=\sum_{j=1}^{k}\frac{x^{b_j}}{1-x^{a_j}},$$

where $k\geqslant 2$ and all the a_j are distinct. Obtain a contradiction by making exactly 1 term $\to\infty$.

91. The generating function form of the problem is $f^2(x)+f(x^2)=P(x)+c/(1-x)$ where $f(x)=\Sigma x^{\alpha}$, $c>0$, and $P(x)$ is a polynomial. Obtain a contradiction by letting $x\to(-1)^{+}$.

92. The generating function form of the problem is

$$\Sigma x^{a}\cdot\Sigma x^{-a}=n-1+\frac{x^{N+1}-x^{-N}}{x-1},$$

where

$$N=n(n-1)/2.$$

Set $x=e^{i\theta}$ and seek a contradiction by making the right side negative.

93. The generating function form of the problem is the identity

$$\frac{1}{1-x}\cdot\frac{1}{1-x^{3}}\cdot\frac{1}{1-x^{5}}\cdots\equiv(1+x)(1+x^{2})(1+x^{3})\cdots.$$

Equivalently $(1-x)(1-x^{3})\cdots(1+x)(1+x^{2})\cdots\equiv 1$, and note that *this* product is unaltered by $x\to x^{2}$.

94. Change variables by $x=\sqrt{nt}$ and apply the dominated convergence theorem.

95. Same as problem 94; set $x=\sqrt{nt}$ and use dominated convergence.

96. Use the integral formula for the Taylor Series expansion.

97. Change variables by $x=y/\sqrt[4]{n}$ and use dominated convergence.

99. Write the recurrence formula for the "failure probabilities," solve it, and then sum them all.

100. The failure probabilities are given by the definite integrals

$$f_n = \underset{\substack{x_i > 0 \\ x_1 + \cdots x_n = 1}}{\int \cdots \int} dx_1\, dx_2 \cdots dx_n$$

which we can evaluate as $1/n!$. Thus, the expectation is e.

101. The failure probabilities are

$$1 \cdot \frac{n-1}{n} \cdot \frac{n-2}{n} \cdot \cdots \cdot \frac{n-k+1}{n} = \binom{n}{k} \frac{k!}{n^k},$$

so that the expectation is

$$\sum_k \binom{n}{k} \frac{k!}{n^k} = \int_0^\infty \left(1 + \frac{x}{n}\right)^n e^{-x}\, dx.$$

Invoke Problem 94.

102. The failure probabilities are $1-(1-1/2^k)^n$ so that the expectation is $\Sigma_{k=0}^{\infty} 1-(1-1/2^k)^n$, and this can be replaced by the definite integral.

103. The failure probabilities must all be positive. Hence $f_n \geqslant 1/2^n$ and so the expectation must be at least $\Sigma_0^\infty 1/2^n = 2$.

105. Most numbers a have around $\log\log a$ and hence $\log\log n$ prime factors, and similarly for b. Hence most $a \cdot b$ have around $2\log\log n$ prime factors. But *most* numbers up to n^2 have only around $\log\log n$ such.

106. Use a category argument to show that $f(x+a)-f(x)$ is bounded for a in some interval $[\alpha, \beta]$, and then pick $h(x) = (1/(\beta-\alpha)) \int_\alpha^\beta f(x+a)\, da$.

107. Use a category argument to make $|f(na)| \leqslant \epsilon$ throughout an a interval $[\alpha, \beta]$ for all $n \geqslant N$. Conclude that $|f(x)| \leqslant \epsilon$ for x large enough.

108. If $f(x)$ is c.m., then so is $f(x+a)$ *and* $f(x)-f(x+a)$ (sic). Thus, for extreme points we must have $f(x+a) = c_a f(x)$, etc.

109. Take an extreme point of such functions (assignments). Notice that it is a positive combination of its own translates. Conclude that it is equal to its translates. Conclusion: Every extreme point is constant. Conclusion: Every one of them is a constant!

SOLUTIONS

1. First we simplify our task by observing that (1) we already have $-$, (2) the $+$ is then obtained by $u+v=u-(0-v)$. Some experimentation with partial fractions leads to the identity

$$\frac{1}{x(1-x)}=\frac{1}{1-x}+\frac{1}{x}.$$

This can then be rewritten as

$$\frac{1}{x-x^2}=\frac{1}{1-x}-\frac{1}{0-x}$$

which gives $1/(x-x^2)$ in terms of $-$ and $1/$. Parlaying this further gives $x-x^2$ and, therefore, x^2. We need then only derive multiplication from "squaring," and this is accomplished by the cliché called polarization. Namely, we have $-2xy=(x-y)^2-x^2-y^2$ which completes our task except for the job of dividing out the -2. This in turn can be accomplished in many ways, e.g.,

$$\frac{u}{-2}=1\bigg/\frac{-2}{u}=1\bigg/\left(\left(0-\frac{1}{u}\right)-\frac{1}{u}\right).$$

2. Building on the previous problem we see that it is sufficient to produce an operation, $\circ$, from which $-$ and $1/$ are derivable. This is also a necessity (since $-$ and $1/$ are among the operations desired). We now try to "picture" such an operation—*literally*—as a formula. We see that something like $x \circ y = x \cdot (x + y)$ could never work. This is composed of the operations of multiplication and addition and could never be iterated to produce, for example, a minus sign! Similarly, a $\div$ sign can never result from such repeated operations. The only time the minus sign or the division sign will emerge is if it is there already! Thus, although $x \cdot (x + y)$ has no chance, something like $1/(x - y)$ or $x - 1/y$ does. So we give it a try. From $1/(x - y)$ we certainly obtain $1/x = 1/(x - 0)$. But also we obtain $x - y = 1/(1/(x - y) - 0)$. So, *success*!

3. The kind of "savings" suggested by the wording of this problem is exemplified by the identity $a \cdot x + a \cdot y = a \cdot (x + y)$. Here two multiplications *seem* to be needed for the left-hand side, whereas only one multiplication is really necessary (as shown by the right-hand side). This example illustrates the possibility of a saving, but is itself not directly useful to us since $ax + ay$ is not an expression of much use for the complex multiplication. *But* $(a + b) \cdot (x + y) = ax + ay + bx + by$ does actually buy us something we can use! Namely, it contains the desired $ay + bx$. If we can also produce ax and by we will have the whole game! Namely, the three products, $\mathrm{I} = (a + b) \cdot (x + y)$, $\mathrm{II} = a \cdot x$, $\mathrm{III} = b \cdot y$, give us $(a + bi)(x + yi) = \mathrm{II} - \mathrm{III} + (\mathrm{I} - \mathrm{II} - \mathrm{III})i$.

Now, what about doing with only 2 products? Experimentation suggests that this is impossible, but how does one *prove* such an impossibility? By hypothesizing two such products and deriving a contradiction. So suppose IV and V are these two hypothetical products. Then there are integers j, k, l, m such that, identically, $ax - by = j \cdot \mathrm{IV} + k\mathrm{V}$, $ay + bx = l\mathrm{IV} + m\mathrm{V}$. How to proceed to a contradiction? One promising direction could be to specialize the variables a, b, x, y so as to make, e.g.,

$ax - by$ equal to 0. So make $a = y, b = x$ and obtain $j\text{IV} + k\text{V} \equiv 0$ so there is only one product now, IV being a rational multiple of V. Thus, $ax + by = r\text{V}$, r being rational, when $a = x$ and $b = y$. Is this the sought-after contradiction? Indeed it is, for it says that $x^2 + y^2 = r \cdot \text{V}$, which means that $x^2 + y^2$ is rationally factored!

4. We can (almost) solve this problem by expressing the unknown probability, x, in terms of itself. For, after all, how can the one microbe multiply forever? Only in one way. It must at the first moment split, and then at least one of its two daughters must succeed in the task of the mother (i.e., lead to an eternal progeny). In short $x = p \cdot (1-(1-x)^2)$. We can solve for x and get the answer! (Almost!) The trouble is that the equation is a quadratic and so has *two* solutions. These are $x = 0$ and $2 - 1/p$. Of course if $p \leqslant \frac{1}{2}$, $2 - 1/p \leqslant 0$, so there is no ambiguity, 0 being the only acceptable probability. But what if $p > \frac{1}{2}$? Then there are two real possibilities, both 0 and $2 - 1/p$. Which is correct? Another idea is needed; namely, the notion of p_n, the probability that our colony last at least n generations. So we have, by the same reasoning as before, $p_{n+1} = p(1-(1-p_n)^2)$, and we are asking about the limit of the p_n. Is it 0 or $2 - 1/p$? Now from the very definition of the p_n it is clear that they are non-increasing, so the question really boils down to whether p_n remains above $2 - 1/p$ (if so the limit must be $2 - 1/p$, and if not it must be 0). But this is a question which can be resolved by a mere induction argument! So assume $p_n > 2 - 1/p$ and deduce, in turn, that $1 - p_n < 1/p - 1$, $(1-p_n)^2 < 1/p^2 - 2/p + 1$, $1-(1-p_n)^2 > 2/p - 1/p^2$, $p(1-(1-p_n)^2) > 2 - 1/p$, and $p_{n+1} > 2 - 1/p$. The induction proof is completed by the observation that $p_0 = 1 > 2 - 1/p$. So in general our solution is $(2 - 1/p)_+$.

Second Solution: There is another pathway to this resolution of the ambiguity, and this has more general applications. So let us suppose quite generally that we have a recurrence formula $x_{n+1}=f(x_n)$, f some smooth function. It is clear that for such a sequence to converge to a limit, say L, we must have $L=f(L)$, which is to say L has to be a fixed point of $f(x)$. But there is another necessary condition for such convergence. A fixed point, L, of $f(x)$ is termed repulsive if $|f'(L)|>1$, and the fact is that x_n cannot approach a repulsive fixed point unless it actually hits it on the nose, i.e., unless $x_n \equiv L$ from some point on. If we delay the proof of this fact for the moment, we can immediately see how it applies to our microbe problem. Here $p_{n+1}=p(1-(1-p_n)^2)$, i.e., $f(x)=p(1-(1-x)^2)$, $f'(x)=p(2-2x)$, $f'(0)=2p>1$ and 0 is a repulsive fixed point, etc.

Finally, let us prove that a repulsive fixed point is never approached (non-trivially). Thus, suppose that $x_{n+1}=f(x_n)$, $L=f(L)$, $|f'(L)|>1$. If any $x_k=L$, then $x_n \equiv L$ for $n \geqslant k$, and we have the "trivial" approach to L; otherwise if $x_n \to L$ we have

$$\frac{f(x_n)-f(L)}{x_n-L} \to f'(L)$$

and so

$$\left|\frac{x_{n+1}-L}{x_n-L}\right| = \left|\frac{f(x_n)-f(L)}{x_n-L}\right| > 1$$

for $n \geqslant N$. But this means that $|x_{n+1}-L|>|x_n-L|$ for all $n \geqslant N$, and so contradicts that $x_n \to L$.

5. As is often the case, it pays to begin with a question—the obvious one—why π/n? For $n=3$ the picture is that of a triangle and the result states that one of the angles of an arbitrary triangle is $\leqslant \pi/3$. Always true, of course, since the sum of the three angles is always π, but also best possible for the equilateral triangle. Similarly for $n=4$, the $\pi/4$ occurs for the square, and generally π/n occurs for the regular n-gon. Namely, each internal angle is $\pi(n-2)/n$, and if we picture all

the diagonals emanating from this angle there are $n-3$ such, and so $n-2$ (equal) angles, all equal to

$$\frac{1}{n-2}\left(\pi\frac{n-2}{n}\right)=\frac{\pi}{n}.$$

This picture of the regular n-gon, however special, does suggest a general argument. Indeed if the n points form any convex n-gon somewhat the same conclusions can be drawn. The sum of all the n vertex angles is still $\pi(n-2)$ and so one, at least, is $\leqslant \pi(n-2)/n$. If from this vertex angle the diagonals are all drawn we have, again, $n-2$ angles and since they add up to $\leqslant \pi(n-2)/n$, one at least is

$$\leqslant \frac{1}{n-2}\left(\pi\frac{n-2}{n}\right)=\frac{\pi}{n}.$$

All we need now fill in is the non-convex case, and this means we have a convex k-gon $(k<n)$ and $n-k$ interior points. But this offers no more troubles than the other case! Choose a vertex of the k-gon where the angle is $\leqslant \pi(k-2)/k$. This is surely $\leqslant \pi(n-2)/n$, and there are still $n-3$ lines emanating to other points (they are not diagonals now, but so what?) We will still obtain our desired angle which is, then,

$$\leqslant \frac{1}{n-2}\left(\pi\frac{n-2}{n}\right)=\frac{\pi}{n}.$$

6. It should be clear that the *real numbers* play a very small role here. All that counts is that we have items from a completely ordered system. The problem gets solved post haste if we introduce the concept of a "giant," namely, an element of the sequence which is *larger* than all the elements that follow it.

   Case 1: There are infinitely many giants. In this case, just take the subsequence consisting of these giants. It is clearly decreasing.

   Case 2: There are only a finite number of giants. Here we start our subsequence with the element after the last giant and we continue by choosing a subsequent element which is not smaller than it and then a next one which is not smaller than this one, etc. This process will continue forever because none of the elements encountered is a giant! We thereby produce a *non-decreasing* subsequence.

7. If we denote $\underbrace{f(f\cdots f(x)\cdots)}_{k}$ by y_k then we have the recurrence relation $y_{k+1} = y_k + (1/n)y_k^2$, $y_0 = x$. But this recurrence rings a bell! Indeed when solving a differential equation $Y' = \phi(\lambda, Y)$, $Y(0) = C$, one uses the *finite difference approach*

$$Y\left(\frac{k+1}{n}\right) = Y\left(\frac{k}{n}\right) + \frac{1}{n}\phi\left(\frac{k}{n}, Y\left(\frac{k}{n}\right)\right), \qquad y(0) = C,$$

and standard theorems state that, as $n \to \infty$, $y_{[nt]} \to Y(t)$. For our case $y_{k+1} = y_k + (1/n)y_k^2$, $y_0 = x$, so we are converging to the solution of $Y'(t) = Y^2(t)$, $Y(0) = x$, which is given explicitly by $Y(t) = x/(1 - xt)$. Since we asked about y_n, we conclude that its limiting value is $Y(n/n) = Y(1) = x/(1-x)$.

Second Solution: Now that we possess the answer, we can see our way clear to another path entirely. In a way, this could prove much simpler and more direct, independent of the theory of differential equations. So let us just be guided by what we have seen, but pretend to begin again.

We consider

$$g(x) = \frac{x}{1 - \dfrac{x}{n}}$$

and ask for $\underbrace{g(g(\cdots g(x)\cdots))}_{n}$. We find that

$$\underbrace{g(g(\cdots g(x)\cdots))}_{k} = \frac{x}{1 - \dfrac{k}{n}x}$$

by induction, since

$$\frac{\dfrac{x}{1-(k/n)x}}{1-\dfrac{1}{n}\dfrac{x}{1-(k/n)x}}=\frac{x}{1-((k+1)/n)x}.$$

Thus our answer is $\underbrace{g(g\cdots g}_{n}(x)\cdots)=x/(1-x)$ and the solution is completed by the observation that $g(x)=f(x)+O(1/n^2)$ so that $\underbrace{g(g\cdots g}_{n}(x)\cdots)=\underbrace{f(f\cdots f}_{n}(x)\cdots)+O(1/n)$.

8. The probability $\frac{1}{3}$ is special: it means we want an experiment whose success probability is half its failure probability. Suppose we keep tossing the coin until a head first shows, and we call it a success if this occurs at an even numbered toss. Clearly then, success is equivalent to the first toss being a *tail* and then the ensuing history being a failure. Thus, success probability $=$ $\frac{1}{2}\cdot$ failure probability, as we desired.

As for doing the same thing for a general p replacing $\frac{1}{3}$, this analysis does not seem to help at all. We have been too slick. This is an ever-present danger: the very reason that a certain solution to a problem is so neat, cute, pretty, or whatever, is that it is *special*, and thus allows no generalization!

There is a more cumbersome way of computing the probability $\frac{1}{3}$ for the success of the experiment just described. Namely, success occurs iff either the first head is on toss number 2 or toss number 4 or toss number 6, etc. These being mutually exclusive, the total probability is $1/2^2+1/2^4+1/2^6+\cdots$ which is equal to $\frac{1}{3}$. The generalization is clear. Let $p=1/2^{n_1}+1/2^{n_2}+\cdots$ be the binary expansion for p, and define success by the condition that the first head appears at the toss numbered n_1 or n_2 or .... This clearly does the job!

How often it happens that the slick and beautiful approach is limited by its own cleverness, while the straightforward humdrum attack wins out in the general case.

9. The guess, of course, is that you can, transcendentals being so much more numerous than algebraics. But how? The hint is the phrase "more *numerous* than algebraics." The algebraics are denumerable! Enumerate them, then, and following Cantor, change the first one in the first place, the second one in (*not the second* but) the third place, etc. (the 5th, 7th, ...). These are your moves, and notice that no matter what your opponent does in the 2nd, 4th, 6th, etc., places, the resulting number is different from all the algebraics. (Of course there is a bit of trouble because of the duplicity of representations, but this really is only a "bit" of trouble and easily repaired.)

10. The point is that half the time, i.e., with probability $\frac{1}{2}$, nothing is lost because the light is green. The other half the time anything from 0 to 30 seconds is lost and by symmetry this averages to 15 seconds. All in all, then, the expected loss is $\frac{1}{2} \cdot 0 + \frac{1}{2} \cdot 15 = 15/2$ seconds.

11. We look at the area of such a hypothetical triangle in two different ways. First, from the determinant formula for a triangle we see that any triangle with lattice point vertices has area $= \frac{1}{2}$ (integer). Second, from the equilateral triangle formula, we have: area $= (\sqrt{3}/4)\text{side}^2 = (\sqrt{3}/4)$ (integer). Comparing these two formulas for the area leads to the fallacy $\sqrt{3} =$ rational!

Second Solution: If we put one vertex of the hypothesized triangle at the origin, the other two at (a, b) and (x, y), then the equilateral condition becomes $a^2 + b^2 = x^2 + y^2 = 2(ax + by)$. The contradiction comes from a simple parity argument. By continually dividing out by two we get to a point where at least one of a, b, x, y is odd. If *wlog* a is odd then, since $a^2 + b^2$ is even, b is also odd and so $a^2 + b^2$ is of the form $8k + 2$. Hence, since $x^2 + y^2 = a^2 + b^2$, x and y are also odd. But then $2(ax + by)$ is not of the form $8k + 2$. Contradiction!

If this second proof seems harder than the first, note that it did not require the knowledge that $\sqrt{3}$ was irrational. Thus, in a real sense, it is actually the simpler proof. As for the three-dimensional question, the answer switches! The vertices $(1,0,0)$, $(0,1,0)$, $(0,0,1)$ obviously span an equilateral triangle.

12. This is obviously a generalization of the fact that a decreasing sequence of positive numbers is convergent. But the structure of the hypothesis has changed in that we are not now comparing a term with a previous term but with an average of them. How to restore the structure and at what price? We are given $x_{n+1} \leqslant (x_n + x_{n-1})/2$; if we choose, e.g., $y_n = (x_n + x_{n-1})/2$, do we obtain $y_{n+1} \leqslant y_n$? *NO*. Try again; this time put $y_n = \mathrm{Max}(x_n, x_{n-1})$ and now the answer is *YES*. ($x_{n+1} \leqslant (x_n + x_{n-1})/2 \leqslant y_n$ and $x_n \leqslant y_n$ so that $y_{n+1} = \mathrm{Max}(x_n, x_{n+1}) \leqslant y_n$). Then $y_n \to L$, but the price is that this just tells us that "half" the sequence converges. But in fact this "half" carries the whole sequence. For y_n, y_{n+1} are both near L, so if $x_n = y_n$ or $x_n = y_{n+1}$ we have the result. Otherwise $x_{n-1} = y_n$ and $x_{n+1} = y_{n+1}$, so $x_n \geqslant 2x_{n+1} - x_{n-1} = 2y_{n+1} - y_n$ and (since $x_n \leqslant y_n$ as noted above) we still have the result.

Second Solution: Here we notice that $x_{n+1} + \frac{1}{2}x_n \leqslant x_n + \frac{1}{2}x_{n-1}$ so that the choice $z_n = x_n + \frac{1}{2}x_{n-1}$ leads to a monotonic and hence convergent sequence. But we can invert this relation and obtain

$$x_n = z_n - \frac{1}{2}z_{n-1} + \frac{1}{4}z_{n-2} + \cdots + \left(-\frac{1}{2}\right)^{n-1} z_1 + \left(-\frac{-1}{2}\right)^n x_0.$$

Using this formula it is easy to read off the convergence of the x_n. Thus, we know that the z_n have a limit L, so replacing all of them by L leaves an error of

$$\epsilon + \frac{\epsilon}{2} + \frac{\epsilon}{4} + \cdots + \frac{\epsilon}{2^{n-N}} + \frac{|z_{N-1} - L|}{2^{n-N+1}} + \cdots < 3\epsilon,$$

and we are led to the conclusion that $x_n \to L(1 - \frac{1}{2} + \frac{1}{4} - \cdots) = \frac{2}{3}L$.

13. This is a case of a recurrence relation with an explicit, closed form, solution. Of course, in such lucky cases convergence questions are fairly transparent. Short of this explicit solution one might be hard pressed for answers.

The point is that

$$\frac{1}{2}\left(\frac{1+z}{1-z} + \frac{1-z}{1+z}\right) = \frac{1+z^2}{1-z^2}$$

so that if we change variables by $x_n = (1+z_n)/(1-z_n)$ our recurrence formula becomes, quite simply, $z_{n+1} = z_n^2$, $z_0 = (1-x_0)/(1+x_0)$. The solution is, of course, $z_n = z_0^{2^n}$, and we can read off the behavior as $n \to \infty$. Thus, if $|z_0| < 1$, $z_n \to 0$; if $|z_0| > 1$, $z_n \to \infty$. If $|z_0| = 1$, there are two cases, namely, with $z_0 = e^{2\pi i t}$, t either is or is not a binary rational (i.e., $a/2^k$, a, k integers). If it is, then $z_n \to 1$; if it is not, then z_n diverges.

Translating back to x_n these results become the following:

If Re $x_0 > 0$, then $x_n \to 1$.
If Re $x_0 < 0$, then $x_n \to -1$.
If Re $x_0 = 0$, then x_n diverges.

What is lost in the translation are the two portions of this last case, namely, becoming equal to ∞ (when $x_0 = i\tan(\pi a/2^k)$ a, k integers) or ordinary divergence (when x_0 is not of this form).

Now that we know the answer we should, perhaps, be able to find a method which does not require the explicit solution. Indeed there is one but it is a real mess. So, a *challenge*: find a *neat* method which avoids the explicit solution!

14. This recurrence formula is satisfied by the constant 1 and by $(-\frac{1}{2})^n$, and so also by any linear combination thereof, namely, $A+B(-\frac{1}{2})^n$. Choosing A, B to fit x_0 and x_1 gives $A=(x_0+2x_1)/3$, $B=(2x_0-2x_1)/3$, so that we have

$$x_n = \frac{x_0+2x_1}{3} + \frac{2x_0-2x_1}{3}\left(-\frac{1}{2}\right)^n$$

and

$$\lim x_n = \frac{x_0+2x_1}{3}.$$

Second Solution: Calling the limit $L(x_0, x_1)$, and noting that this same limit is approached starting with x_1, x_2 instead, gives the functional equation $L(x_0,x_1)=L(x_1,(x_0+x_1)/2)$. If we also note that L is linear and $L(1,1)=1$, we obtain, as before, $L(x_0,x_1)=(x_0+2x_1)/3$. Note that the recurrence equation $x_{n+1}=\sqrt{x_n x_{n-1}}$ is obtainable from this one by a logarithmic change of variables so that here $\lim x_n = \sqrt[3]{x_0 x_1^2}$.

15. This fails to succumb to standard calculus techniques because of the requirement of being integers. If they were merely positive numbers we could use those techniques, and we would find that the numbers should be taken equal and as nearly equal to e as we can make them. Integers cannot be nearly equal to e. Equal to 3 is the nearest we can do. So, GUESS: split n into 3's, with a 2 if needed or two 2's if that's needed and this will give the biggest product (e.g., for $n=10$ the split is 3, 3, 2, 2 for a product of 36; for $n=12$, it is 3, 3, 3, 3 giving 81; and for $n=20$, it is 3, 3, 3, 3, 3, 3, 2 which gives 1458).

The verification is actually very simple and direct. Suppose we have the best partition of n and we look at a maximum summand, m. If $m>3$, we have $2\cdot(m-2)\geqslant m$, so at no loss we can lower m. Thus, there is an optimal partition made up of 2's and 3's. (1's are clearly worthless.) Next we observe that there cannot be more than two of the 2's for the replacement of three

2's by two 3's is an improvement ($2^3 = 8 < 9 = 3^2$). Our guess is now verified.

16. The one weapon we possess is the arithmetic-geometric (AG) inequality,

$$\sqrt[n]{a_1 a_2 \cdots a_n} \leqslant \frac{a_1 + a_2 + \cdots a_n}{n},$$

and this just is not good enough. It leads to the series

$$a_1 + \frac{a_1 + a_2}{2} + \frac{a_1 + a_2 + a_3}{3} + \cdots$$

which is certainly divergent (being bigger than $a_1(1 + \frac{1}{2} + \frac{1}{3} \cdots)$). We seem to be stymied: the only weapon we have is indeed AG, and it does not work. But this is because we are being very wasteful. AG is only sharp when the terms are almost equal, and the a_n certainly are not. The "border-line" of convergent series is, of course, $1/n$ so perhaps we should be applying AG to na_n instead. This gives

$$\sqrt[n]{a_1 a_2 \cdots a_n} \leqslant \frac{1}{\sqrt[n]{n!}} \frac{a_1 + 2a_2 + \cdots na_n}{n}$$

which leads to the series

$$a_1 + \frac{1}{\sqrt{2}} \frac{a_1 + 2a_2}{2} + \frac{1}{\sqrt[3]{6}} \frac{a_1 + 2a_2 + 3a_3}{3} + \cdots$$

and perhaps this is convergent. So we look further. A crude estimate via Stirling's formula tells us that $\sqrt[n]{n!}$ is of order n, and so the series looks like

$$\Sigma a_n \cdot n\left(\frac{1}{n^2} + \frac{1}{(n+1)^2} + \cdots\right).$$

The tail $1/n^2 + 1/(n+1)^2 + \cdots$ is of order $1/n$, however, so the series looks like Σa_n itself. This works! The problem is hereby solved, but we prefer now to neaten up the details.

So we want a lower bound for $n!$ The simplest good one is $n! > (n/e)^n$; this is not really what we want, but it is so pretty and simple that we felt everyone should see it at least once! Namely, $n^n/n!$ is one of the terms in the expansion of e^n, hence $n^n/n! < e^n$: Q.E.D.! More convenient for us is the inequality $n! > ((n+1)/e)^n$, and this is almost as easy. Namely,

$$\frac{(n+1)^n}{n!},$$

$$\frac{(n+1)^{n+1}}{(n+1)!} = \frac{(n+1)^n}{n!},$$

and

$$\frac{(n+1)^{n-1}}{(n-1)!} = \frac{n}{n+1}\,\frac{(n+1)^n}{n!}$$

are three terms in the expansion of e^{n+1}, and for $n > 2$, $e < 2 + n/(n+1)$, so that

$$\left(2 + \frac{n}{n+1}\right)\frac{(n+1)^n}{n!} < e^{n+1}$$

yields $(n+1)^n/n! < e^n$. The rest is truly neat: since $\sqrt[n]{n!} > (n+1)/e$, we see that our upper bound series is less than

$$e\left(\frac{a_1}{2} + \frac{a_1 + 2a_2}{2\cdot 3} + \frac{a_1 + 2a_2 + 3a_3}{3\cdot 4} + \cdots\right)$$

$$= e\left(a_1\left(\frac{1}{1\cdot 2} + \frac{1}{2\cdot 3} + \cdots\right) + a_2\left(\frac{2}{2\cdot 3} + \frac{2}{3\cdot 4} + \cdots\right) + \cdots\right.$$

and this is, in turn, equal to $e(a_1 + a_2 + a_3 + \cdots)$ since $\sum_{v=k}^{\infty} k/v(v+1) = 1$ for all k.

We have proved that $\sum\sqrt[n]{a_1 a_2 \ldots a_n} < e\sum a_n$; by looking a little closer we can also see that the e in this inequality is best possible. Our inequalities, remember, were designed to be exact for the sequence $1/n$. Since this gives a divergent series it behooves us to look at its truncation. So set $a_n = 1/n$ for $n \leqslant N$ and 0 thereafter. In this case $\Sigma a_n \sim \log N$, while $\sqrt[n]{a_1 a_2 \ldots a_n} = 1/\sqrt[n]{n!} \sim e/n$, by Stirling's formula. Thus, $\Sigma\sqrt[n]{a_1 a_2 \ldots a_n} \sim e \log N$, so the e cannot be improved upon.

17. The first thing one thinks of in this regard is Fermat's theorem that, for p prime, $x^p - x \equiv 0 \pmod p$. This degree is furthermore minimal since, in a field, a kth degree polynomial has at most k zeros. In short, we have the solution when n is prime.

But this turns out to be a false lead. The primes are altogether too special. More to the point is the very basic question: when is a polynomial integer valued on the integers? (This is just what we are asking of $1/n$ (our monic polynomial).) The fact is, of course, that it is not necessary for an integer-valued polynomial to have integer coefficients (just think of $(x^2 + x)/2$). The answer to this question is well known: a polynomial is integer valued if and only if it is an integer combination of the polynomials $\binom{x}{k}$ $(= x(x-1)\cdots(x-k+1)/k!)$, for every polynomial can be written as $c_n\binom{x}{n} + c_{n-1}\binom{x}{n-1} + \cdots + c_0$. If this polynomial is integer valued, then set $x = 0$ and deduce that c_0 is an integer. Next set $x = 1$ and obtain that $c_1 + c_0$ is an integer, so that c_1 is. In general, setting $x = k$ gives the expression $c_k +$ integer combination of earlier ones and the result is established inductively.

In our case, then, we must have $1/n$ (our monic polynomial) $= \Sigma_{k=0}^{m} a_n \binom{x}{k}$, a_k integers, so equating leading coefficients gives $1/n = a_m/m!$ which means that $n|m!$. Conversely, if $n|m!$, then the choice $a_m = m!/n$, $a_k = 0$ for $k < m$ gives us a monic polynomial of the sort required.

So the answer is: this minimal degree is the smallest m for which $n|m!$. In particular for $n = 100$, m is 10, the polynomial being $x(x-1)(x-2)\cdots(x-9)$.

18. Finding a single number is always a hard job. We do better on functions—they have a certain built-in context. So consider instead

$$f(x)=\sqrt{1+x\sqrt{1+(x+1)\sqrt{1+\cdots}}}\,,$$

which satisfies the functional equation $f^2(x)=1+xf(x+1)$. One solution, we observe, is the function $x+1$: $((x+1)^2=1+x(x+2))$. But is this the only solution? Is it *our* $f(x)$? This is a difficult question, but the "guess" $x+1$ does suggest a tack, namely, let us bound above and below by constant multiples of $x+1$. Below: if we strip away all the + signs we obtain the lower bound of

$$\sqrt{x\sqrt{x\sqrt{x\cdots}}}=x\geqslant\tfrac{1}{2}(x+1)\qquad(\text{for } x\geqslant 1).$$

Above: here we beef up all the terms to read

$$\sqrt{(x+1)\sqrt{(x+2)\sqrt{(x+3)\cdots}}}\,,$$

and then even more beefed up it reads

$$\sqrt{(x+1)\sqrt{2(x+1)\sqrt{4(x+1)\cdots}}}=A\cdot(x+1),$$

$$A=\sqrt{1\sqrt{2\sqrt{4\cdots}}}=2.$$

So $\frac{1}{2}(x+1)<f(x)<2(x+1)$, but then these bounds improve! We have $\frac{1}{2}(x+2)<f(x+1)<2(x+2)$ so that, since $f^2(x)=1+xf(x+1)$, $\frac{1}{2}+xf(x+1)<f^2(x)<2+xf(x+1)$, $\frac{1}{2}(1+x(x+2))<f^2(x)<2(1+x(x+2))$ and $\sqrt{\frac{1}{2}}\,(x+1)<f(x)<\sqrt{2}\,(x+1)$. Repeating gives $\sqrt[2^k]{\frac{1}{2}}\,(x+1)<f(x)<\sqrt[2^k]{2}\,(x+1)$, and letting $k\to\infty$ shows $x+1\leqslant f(x)\leqslant x+1$. Indeed $f(x)=x+1$, and our numerical answer is 3.

19. Associate with each person the number of friends he has at the party. Thus, if there are n people, each one is associated with a number from 0 to $n-1$. But it is impossible that the 0 and the $n-1$ both appear. Hence two must be equal. (Note that three assumptions were tacit here: friendship is reflexive, no one counts himself as a friend, and a party consists of at least 2 people.)

20. We are certainly reminded of Fermat's theorem that, for p a prime, $p|(2^{p-1}-1)$ (so that, of course, $p \nmid (2^p-1)$). Perhaps less familiar is a kind of converse, to wit, that if $p|(2^n-1)$ then n has a non-trivial factor in common with $p-1$. For assume the contrary, that $(n, p-1)=1$. Then, for some integers x and y, $xn+y(p-1)=1$, and this leads to the contradiction $2=(2^n)^x \cdot (2^{p-1})^y \equiv 1 \pmod{p}$.

Now, as we saw, Fermat's theorem settles our problem when n is a prime. But in fact this converse theorem settles it in all cases. Just choose p as the *smallest* prime divisor of n, and notice that then $p-1$ has no primes in common with n. Conclusion: $p \nmid (2^n-1)$, and so, of course, $n \nmid (2^n-1)$.

21. Since there are $2\cdot 3^{n-1}$ non-multiples of 3 in the $(\bmod 3^n)$ system we see that we are being asked to prove that the powers of 2, namely, 2, 4,... $2^{2\cdot 3^{n-1}}$ $(\equiv 1)$, are all incongruent $(\bmod\ 3^n)$. If 2^i were congruent to 2^j, with $j>i$ say, then $2^{j-i}\equiv 1 \pmod{3^n}$. By the induction hypothesis, however, since $2^{j-i}\equiv 1 \pmod{3^{n-1}}$ we could conclude that $j-i=2\cdot 3^{n-2}$, or $4^{3^{n-2}}\equiv 1 \pmod{3^n}$. But we see instead, by induction, that $4^{3^{n-2}}\equiv 1+3^{n-1} \pmod{3^n}$. Namely, from $4^{3^{n-2}}=1+3^{n-1}+M\cdot 3^n$, it follows that $4^{3^{n-1}}=1+3^n+M\cdot 3^{n+1}+3(3^{n-1}+M3^n)^2+(3^{n-1}+M3^n)^3$, and the last two terms are clearly seen to be multiples of 3^{n+1}.

22. (The problem would be much simpler (mod p^n) for p an odd prime, and this is some more validation of the old belief that 2 is the *oddest* prime of all!)

We simplify matters by looking first at the odd squares (mod 2^n), it being fairly clear how we can get the full tally from this one. The perverse nature of 2 is evident when we observe that this number stays 1 for $n = 1, 2, 3$. However, from there on, we easily see, the number doubles each time n increases by 1. Namely, if x is odd, then $(x + 2^{n-1})^2 = x^2 + 2^n x + 2^{2n-2} \equiv x^2 + 2^n \pmod{2^{n+1}}$ (since we are looking at $n \geqslant 3$ so that $2n - 2 \geqslant n + 1$). And so we see that each odd square, $x^2 \pmod{2^n}$ becomes two odd squares, x^2 and $x^2 + 2^n \pmod{2^{n+1}}$.

Thus, the number of odd squares (mod 2^n) is 2^{n-3} for $n \geqslant 3$. As we promised, we can now write down the total number of squares (mod 2^n), for these are either odd, (numbering 2^{n-3}) or 4 times an odd (numbering $2^{(n-2)-3}$) or 4^2 times an odd (numbering $2^{(n-4)-3}$), etc. The grand total, taking into account the singular behavior at the beginning and adding on one more for the square 0, comes to $1 + 2^{n-3} + 2^{n-5} + \cdots + 2^0 + 2^0$ for n odd, $1 + 2^{n-3} + 2^{n-5} + \cdots + 2^1 + 2^0$ for n even. Summing these geometric series gives, then, $(2^{n-1} + 5)/3$ for n odd, $(2^{n-1} + 4)/3$ for n even. (Roughly speaking about one-sixth of the numbers are perfect squares.)

23. At 0, 1, and ∞ the function is equal to 1. If we sample some other points, e.g., 2, we find it is smaller: $\frac{1}{4} + \sqrt{\frac{1}{2}} = 0.957\ldots$. So we guess that the answer is equal to 1, giving the charming inequality $2^{-x} + 2^{-1/x} \leqslant 1$ on $(0, \infty)$. (At any rate it helps to know the answer in advance.)

The trouble with the Calculus in general, and with using it in this case in particular, is that it leads to unsolvable transcendental equations. Namely, we obtain $2^{-x} = 2^{-1/x} \cdot 1/x^2$. This does have the solution $x = 1$, but it has other solutions as well and we cannot tell which is the maximum point.

This is the way out: at the maximum point we have $2^{-1/x} = x^2 2^{-x}$, so let us plug this into our desired inequality and see if that new inequality is any easier. Indeed the new inequality looks simpler, namely, it is $2^{-x} + 2^{-x}x^2 \leqslant 1$ or $1 + x^2 \leqslant 2^x$. Yes, this looks simpler; the only trouble is that it is false. ($1+3^2 > 2^3$; $1+4^2 > 2^4$, etc.) But all is not lost, these examples have $x > 1$, perhaps $1 + x^2 \leqslant 2^x$ does hold on $[0, 1]$. If so, we still win the game since our original inequality $2^{-x} + 2^{-1/x} \leqslant 1$ is symmetric in x, $1/x$ and so need only be established on $[0, 1]$. The easiest way to see that $1 + x^2 \leqslant 2^x$ on $[0, 1]$ is by noticing, by direct differentiation, that $1 + x^2 - 2^x$ is convex on $[0, 1]$, a convex function always taking its maximum at an endpoint of an interval.

24. This is *not* a geometry problem at all but one in pure combinatorics. We did not specify *where* (i.e., from what space) these points were given, and indeed the only "axiom" which was meant was that any two distinct points determine a line. (Perhaps, then, we should restate the problem in its pure form where the lines are elements and the points are sets of elements, namely, those lines which contain the point.)

So, restatement: Given n distinct sets $S_1, S_2, \ldots, S_n$ such that, for any i, j with $i \neq j$, $|S_i \cap S_j| = 1$, show that $|\bigcup_{i=1}^{n} S_i| \geqslant n$.

Be that as it may, we now proceed by, of all things, linear algebra! Namely, if we assume "par contraire" that there are more points than lines, we can attach numbers, not all 0, to the points in such a way that for each line the total of these numbers for its points is equal to 0. So call the points $1, 2, 3, \ldots, n$, the lines $L_1, L_2, \ldots, L_k$, and the numbers $x_1, x_2, \ldots, x_n$. We have, then, $\Sigma_{i \in L_v} x_i = 0$ for each v. Now look at $\Sigma_{v=1}^{k}(\Sigma_{i \in L_v} x_i)^2$, and observe that every x_i^2 appears *at least twice* and that every cross term $2x_i x_j$, $i \neq j$ appears exactly *once*. Thus,

$$0 = \sum_{v=1}^{k} \left(\sum_{i \in L_v} x_i \right)^2 \geqslant \sum_{i=1}^{n} x_i^2 + \left(\sum_{i=1}^{n} x_i \right)^2 > \left(\sum_{i=1}^{n} x_i \right)^2 \geqslant 0,$$

the sought-for contradiction!

25. Suppose we determine x_1 such that $x_1 \cdot f(x_1) = 1/n^2$ and then x_2 so that $(x_2 - x_1)\ (f(x_2)) - (f(x_1)) = 1/n^2$ and then x_3 so that $(x_3 - x_2)\ (f(x_3) - f(x_2)) = 1/n^2$, etc. We can do this up through x_{n-1} by continuity—even if f itself is not continuous, by the convention of deeming $f(x)$ as multivalued, i.e., having *all* values between $f(x^-)$ and $f(x^+)$. The burden is to then show that the remaining rectangle has area $\leqslant 1/n^2$, i.e., that $(1 - x_{n-1})(1 - f(x_{n-1})) \leqslant 1/n^2$, for then we can expand this last rectangle to have area exactly $1/n^2$.

To this end we have, on the one hand,

$$\begin{aligned}\sqrt{x_1}\sqrt{f(x_1)} &+ \sqrt{x_2 - x_1}\sqrt{f(x_2) - f(x_1)} + \cdots \\ &+ \sqrt{x_{n-1} - x_{n-2}}\sqrt{f(x_{n-1}) - f(x_{n-2})} \\ &+ \sqrt{1 - x_{n-1}}\sqrt{1 - f(x_{n-1})} \\ &= \frac{n-1}{n} + \sqrt{1 - x_{n-1}}\sqrt{1 - f(x_{n-1})},\end{aligned}$$

and on the other hand

$$\begin{aligned}\sqrt{x_1}\sqrt{f(x_1)} + \ldots &\leqslant \sqrt{x_1 + (x_2 - x_1) + \cdots} \\ &\cdot \sqrt{f(x_1) + (f(x_2) - f(x_1)) + \cdots} = 1\end{aligned}$$

by the Schwarz inequality. Comparing these two gives

$$\frac{n-1}{n} + \sqrt{((1 - x_{n-1})(1 - f(x_{n-1}))} \leqslant 1,$$

which implies $(1 - f(x_{n-1})(1 - x_{n-1})) \leqslant 1/n^2$, as we desired.

26. The construction we give will do this covering without "tilting" the squares and then it becomes a very pretty result because the 3 is best possible. Namely, 3 squares each of area $1 - \varepsilon$ cannot cover: they cannot even cover the 4 vertices, since each covers at most one!

Sometimes the "dumbest" approach to a problem is the one that leads to success, and this happens to be the case here. Just arrange the squares in decreasing (non-increasing) order and line them up in rows until the length of each row first exceeds 1. We obtain, thereby, rows of length >1 and *minimum* heights which we may call $h_1, h_2, h_3, \ldots$ and finally a row of length $\leqslant 1$ with a height of perhaps 0. We need only prove that $h_1 + h_2 + h_3 + \cdots \geqslant 1$ for then the juxtaposition of these rows will cover the whole unit square. Now the *maximum* heights of these rows are bounded by $1, h_1, h_2$, etc., and so the areas of said rows are bounded by $1\cdot(1+h_1), h_1(1+h_2), h_2(1+h_3), \ldots$ which are then bounded by $1+h_1, h_1+h_2, h_2+h_3, \ldots$. Comparing areas then gives

$$1+2h_1+2h_2+\cdots \geqslant \text{Total Area} \geqslant 3$$

or $h_1+h_2+\cdots \geqslant 1$, as desired.

27. Since two squares of side $\frac{1}{2}+\epsilon$ cannot be so fit into the unit square, this is also a best possible result.

We proceed as in Problem 26, arranging the squares in decreasing order and then making rows of them. This time, however, we stop each row just *before* its length exceeds 1. So, calling the *maximum* heights $H_1, H_2, \ldots$, we see that this time we want the inequality $H_1 + H_2 + \cdots \leqslant 1$ (so that the "pile up" of all the rows will then fit into the unit square).

To obtain this inequality, we call the lengths of the rows $L_1, L_2, \ldots$, respectively, and note that each $L_k + H_{k+1} > 1$. Hence, $L_k + H_{k+1} - H_k > 1 - H_k \geqslant 1 - H_1$, and so $H_{k+1}(L_k + H_{k+1} - H_k) > (1-H_1)H_{k+1}$. If we sum these inequalities we note that the left-hand sides add up to an area bounded by the full sum of the individual squares *less* the first one; i.e., to $\frac{1}{2} - H_1^2$. So we have $\frac{1}{2} - H_1^2 \geqslant (1-H_1)(H_2 + H_3 + \cdots)$ and this, in turn, gives $\frac{1}{2} + H_1 - 2H_1^2 \geqslant (1-H_1)(H_1 + H_2 + H_3 + \cdots)$. The proof is completed by noting that $\frac{1}{2} + H_1 - 2H_1^2 = 1 - H_1 - \frac{1}{2}(1-2H_1)^2 \leqslant 1 - H_1$; which gives our required inequality, namely, $1 - H_1 \geqslant (1-H_1)(H_1 + H_2 + H_3 + \cdots)$.

28. It might be *very* small. One point? No. Two points, also no. For any two points at distance d, say, we can find a point at any distances α and β as long as $\alpha+\beta > d$ and $|\alpha-\beta| < d$, and there are rationals which do this. But what about three points? The simplest thing to try is three points on the x axis say at $0, 1, \xi$. To be at rational distances from these points, we would need $x^2 + y^2 = \text{rational}$, $(x-1)^2 + y^2 = \text{rational}$, $(x-\xi)^2 + y^2 = \text{rational}$. The first two imply that $2x - 1$, and hence x, is rational. Hence, y^2 is, and so from the third condition ξ has to satisfy a quadratic equation. Conclusion: the three points (on the x axis) $0, 1, \xi$ do the job as long as ξ is neither rational nor a quadratic surd. So, to be specific, $0, 1, \sqrt[3]{2}$ do it.

29. The key observation is that if A and B are in our collection of points, then *all* the points P satisfy $\text{dist}(A, P) - \text{dist}(B, P) =$ integer bounded by $\text{dist}(A, B)$. Thus, all the P lie on a finite number of hyperbolas with axes along $\overline{AB}$. But if there were a third point, C, not in line with A and B, then we could conclude the same thing for $\overline{BC}$. That is, all the P lie on finitely many hyperbolas with axes along $\overline{BC}$.

Putting these two bits together tells us that all our points lie on the intersections of finitely many distinct (having different axes) hyperbolas. This is indeed a finite set!

30. We dispense instantly with the case of even n, they *never* work. For them, $(2a+1)^n + (2b+1)^n$ is twice an odd integer, and so surely not divisible by 2^n. If n is odd, then we have a factorization to work with. Specifically, $(2a+1)^n + (2b+1)^n = 2(a+b+1)((2a+1)^{n-1} - (2a+1)^{n-2}(2b+1) + - \cdots)$. The second factor is a sum of n (an odd number) terms, each of which is odd. Thus, the second factor is odd, and so for 2^n to divide $(2a+1)^n + (2b+1)^n$ it must divide $2(a+b+1)$. This occurs only finitely often.

31. Yes. But one of the difficulties in proving it is the potential unboundedness of the degrees. Luckily the reals are nondenumerable and so, for infinitely many y, $f(x, y)$ has the same degree as a polynomial in x. In other words, for these y, $f(x, y) = f_0(y) + f_1(y)x + \cdots + f_N(y)x^N$ identically in x. Since, for each fixed x (in particular for $N+1$ different values of x), this is a polynomial in y, it follows that the $f_k(y)$ are polynomials. This means that, for each x, $f(x, y) = P(x, y)$, a polynomial, for infinitely many y, but for that value of x $f(x, y)$ is itself a polynomial in y. A polynomial in y equal to $P(x, y)$ for infinitely many y? This must *be* $P(x, y)$.

Conclusion: $f(x, y) \equiv P(x, y)$ for *all* x, y.

We really needed the non-denumerability of the reals. Indeed, if our underlying field were the rationals, say, not only would our proof fail, but there would be a counterexample! Let $r_1, r_2, r_3, \ldots$ enumerate our field, and consider the series $(x - r_1)(y - r_1) + (x - r_1)(x - r_2)(y - r_1)(y - r_2) + \cdots$. This is terminating for each fixed x, y so there is no question of divergence. For each fixed x it is a polynomial in y and vice versa. But, since the y degree is *exactly* $n - 1$ for $x = r_n$, this function is definitely not a polynomial in the joint variables.

Amusing is the fact that, since for finite fields all functions are polynomials, we see that our answer is YES for any field of cardinality not $\aleph_0$ and NO otherwise!

32. It is tempting to say that these three integers are "almost" relatively prime and so their product cannot be a kth power unless each of them is. The catch is the "almost." If n is odd then of $n-1, n, n+1$ the $n-1$ and $n+1$ do share a factor, 2. Why could not $n - 1 = 2x^3$, $n = y^3$, $n + 1 = 4z^3$, say? The correct proof comes from noting that $(n-1)\cdot(n+1)$ and n *are* relatively prime so that they would have to both be kth powers and then the *consecutive* integers $n^2 - 1$ and n^2 would both be kth powers, which is impossible!

33. This problem can really be perplexing if one takes the wrong point of view. This wrong approach being the attempt to *construct* this large disc. The passive (female?) approach is the one that succeeds here. Namely, we assume that there is no disc of radius r lying in the region. Then every point in the region is within r of the boundary. This means that if we take strips of width r along each side they (essentially) cover the region. Some trouble is encountered at the vertices, so we need to do some careful bookkeeping here. Namely, if we have a side from A to B, then we erect not a strip, but the set within r of AB and also within the angle bisectors at A and B. The area of this set is equal to $r \times$ length of $AB +$ two contributions from the endpoints. If the angle at the endpoint, say A, is $< \pi$ then we must subtract the right triangle of height r and angle $A/2$, and this has area bigger than the circular sector it contains, viz., $(r^2/4)(\pi - A)$. If the angle at the endpoint, say B, is $\geqslant \pi$, then we must add on the sector of angle $(B - \pi)/2$, an area of $(r^2/4)(B - \pi)$. In all, then, the total area of these sets is bounded by $r \cdot$ perimeter $-(r^2/2)\Sigma$ exterior angles $=$ $r \cdot$perimeter$- \pi r^2$, and since the union of these sets contains the whole region by hypothesis, we see that Area $\leqslant r \cdot$perimeter$-$ πr^2. This being true for all r in excess of the inscribable radius shows that area/perimeter is a (strict) lower bound for this radius.

34. At first glance 7 seems an awful lot of questions. 4 would suffice if no lie were allowed. 3 questions to cover 1 lie? But yes—7 possibilities plus 1 (no lie), and 8 possibilities require exactly 3 yes or no answers to be resolved. So how?

The numbers 0 through 15 are precisely the 4 (binary) digit numbers. If we ask for the first 3 digits with our first 3 questions, this is a start. Next ask, "Have you lied yet?" An amusing question—but in fact a very decisive one! If the answer is NO, then this is *automatically* correct (else this would be a lie, as well as the one before). You now have 3 questions left to determine the last digit, an easy task. But suppose the answer to this fourth question was YES. This is not automatically correct, *BUT*, it is automatic that the lie has been used up in these first 4 questions.

The lie is used up, and so you find out where by using two questions (these suffice to decide among 4 possibilities), and then you use your last question to find the last digit.

35. We use a continuity argument, albeit a rather elaborate one. So pick a direction, θ, and (by a continuity argument) produce the unique directed line in that direction which bisects the region. Next, through any point on this line we draw another directed line so that the *forward* wedge contains one-sixth of the region. When this point is chosen all the way *back* on the line, then the *backward* wedge contains almost 0 of the region and when the point is all the way forward, this backward wedge contains almost one-half of the region. Thus, by continuity we may choose to make the forward wedge and the backward wedge each contain one-sixth of the region. This is our second line, and for the third line we can, in the same way, arrange for the forward and backward wedges to each contain one-third of the region.

These three lines are not necessarily concurrent (if they were the problem would be solved), but we define $f(\theta)$ as the directed length from the first intersection point (meeting point of the first line and the second line) to the second intersection point (meeting point of the first line and the third line).

Now when θ is replaced by $\theta + \pi$, the entire picture remains the same *except* for the orientation and order of the lines. In short, we obtain $f(\theta + \pi) = -f(\theta)$. Our final continuity argument says that $f(\theta)$, therefore, has a zero, and as we said before, this gives us our desired concurrent lines.

36. Draw the three lines à la Problem 35. The point where they meet obviously has this property since any line through it has two wedges + some more on each side of it.

37. It's a joke. Just write $a = A^2$, $b = B^2, \ldots$, etc., and note what it says is that

$$\sqrt{A^2 + B^2 + C^2 + D^2 + \cdots} + \sqrt{0^2 + B^2 + C^2 + \cdots}$$
$$+ \sqrt{0^2 + 0^2 + C^2 + D^2 + \cdots}$$
$$\geqslant \sqrt{A^2 + (2B)^2 + (3C)^2 + \cdots}\ .$$

This is just the triangle inequality, saying that the sum of the lengths of the vectors $(A, B, C, \ldots)$, $(0, B, C, D, \ldots)$, $(0, 0, C, D, \ldots), \ldots$ is at least the length of the sum of these vectors, namely, $(A, 2B, 3C, 4D, \ldots)$. Simple!

38. If 2 is a perfect square (quadratic residue) mod p, then we are done, and similarly if -2 is a perfect square. The point is that if a and b are both *not* perfect squares, then $a \cdot b$ *is*. Hence we may assume that $-2 \cdot 2 = -4$ is a perfect square, say x^2 (mod p), and we may assume p is odd. But then $(1 + x/2)^2 = 1 + x + x^2/4 \equiv x \pmod{p}$ so that -4 is a perfect 4th power (of $1 + x/2$), and again we are done. (A curious fact emerges here, that if -4 is a perfect square, then it is automatically a perfect fourth power.)

39. Assume not, and look at any unit equilateral triangle. The point is that the vertices must be of all three different colors. So now if we view two unit equilateral triangles which share a face we are forced to the conclusion that the distant vertices ($\sqrt{3}$ apart) must be of the same color. A little thought shows that this means that *any* two points $\sqrt{3}$ apart have the same color. But, therefore, the whole circle of radius $\sqrt{3}$ is unicolored, and surely there are plenty of pairs of points there of mutual distance 1.

40. If not, then the reds miss a distance a, and the blues miss a distance b; *wlog* we may assume $a \leqslant b$. Starting at a blue point

(there must *be* a blue point) we construct an isosceles triangle of sides a, b, b (with the blue point at the vertex of the two b sides); the contradiction is immediate.

41. The baffling part of this construction is that it cannot be made locally. There just seems no controlling how close some later point will approach. Perhaps the best way of seeing how to proceed is by first treating the case of a *polygonal* arc.

Our first step is to remove a small neighborhood of each of the vertices. This leaves a batch of disjoint line segments of total length still in excess of 1. Now *here* is where we choose n (and it *is* done globally as was predicted above). Namely, we make $1/n \leqslant$ the distance between any two of our line segments. Having chosen this n, we do what the conditions dictate us to do. We split each line segment into, say equal, pieces of length at least $1/n$. Thus, if the length of the line segment is L, we break it into $[nL]$ segments, but this gives $[nL]+1 > nL$ endpoints, and since the sum of these L's exceeds 1, the total number of these points exceeds n.

Guided by this special case we proceed to the general one. This time we draw consecutive chords whose total length exceeds 1, and then we separate them by a tiny amount at each endpoint. We have thereby produced disjoint arcs whose chords still have length in excess of 1. We choose n so that $1/n \leqslant$ the distance between any two of these arcs, then we split these chords into lengths $\geqslant 1/n$, and observe that the total number of endpoints exceeds n. All this is as before, but now we need one more step. We project, perpendicularly from these points on the chord, up to the corresponding points on the corresponding arc. These are our desired points. There are indeed more than n of them and they are all at least $1/n$ apart from one another.

42. The 10 and 9 are of course a joke. They might as well be 1 and 0. A weighing tells us just how many good coins there are in the

batch. So suppose we weigh coins 1 and 2. If they are both good or both bad then we are easily done, so let us assume that precisely one of them is good. Now we can weigh coins 1 and 3, and again if these are both good or both bad we will have determined 1, 2, and 3 and we are easily done. So we again assume that precisely one is good. But now we know that 2 and 3 are similar coins, and so by finally weighing 2, 3, and 4 everything gets determined.

Although the problem is thereby solved, there is really more to be said. A curious and perhaps quite general observation can be made. Our analysis above is of a sequential nature. We have described a procedure which has a "branching" structure. (If this and this happens, then do that and that.) This is quite usual in these weighing problems—but is it necessary? The answer, at least in this case, is NO! Let us weigh 1 and 2, then (no matter what) weigh 1 and 3, and then (again no matter what) weigh 2, 3, and 4. These three weighings do indeed determine everything. We have the three equations $x + y = a$, $x + z = b$, $y + z + w = c$, where x, y, z, w are each either 0 or 1. By adding these three equations we obtain $2(x + y + z) + w = a + b + c$, and we may thereby determine w. It is 0 if $a + b + c$ is even, and 1 otherwise. Having thus determined w, we go back and determine x, y, and z from the same three equations. No branching was necessary!

43. The $\frac{1}{6}$ comes from the interval $[\frac{1}{3}, \frac{1}{2}]$ which cannot be multiplied by any n without covering some integer. So the $\frac{1}{6}$ is best possible, but also we see how to get at the positive result. Namely, let us consider the possibilities of doubling or tripling an interval without integers. If doubling introduces an integer, then our interval contains a number $k/2$, k odd. If also tripling introduces an integer, then our interval contains a number $j/3$, j an integer. Since the distance between $j/3$ and $k/2$ is at least $\frac{1}{6}$, we see that our length can always be increased unless it is already at least $\frac{1}{6}$. (Furthermore, we see that our integer n can in fact be chosen as $2^a 3^b$, a, b non-negative integers.)

44. Numerically $(\sqrt{2}+1)^n$ is equal to 2.414..., 5.828..., 14.071..., 33.968..., 82.012..., ... so the integer parts are 2, 5, 14, 33, 82,... and, as far as they go, these *are* alternately even and odd. But we also notice the behavior of the numbers after the decimal point, or rather the distances to the nearest integer. These are 0.414..., 0.171..., 0.071..., 0.031..., 0.012..., ..., a geometric progression? Aren't they in fact $(\sqrt{2}-1)^n$? Indeed, since $1-\sqrt{2}$ is the algebraic conjugate of $\sqrt{2}+1$ we see that $(\sqrt{2}+1)^n+(1-\sqrt{2})^n$ is an integer. A little closer look shows that it is an *even* integer: either note that $\sqrt{2}+1 \equiv \sqrt{2}-1$ (mod 2) as algebraic integers, or consider the binomial expansion

$$(\sqrt{2}+1)^n+(1-\sqrt{2})^n=2\left(1+2\binom{n}{2}+2^2\binom{n}{4}+\cdots\right).$$

At any rate, $-1<1-\sqrt{2}<0$, so our result is established.

45. We try to solve this minimum problem by looking at the continuous one. So note that $x+n/x$ is a unimodal function with its trough at $x=\sqrt{n}$. Thus, the same is true of $[x+n/x]$, so specializing x to be an integer we conclude that the minimum is taken at either $[\sqrt{n}]$ or $[\sqrt{n}]+1$. To decide which let us write $j=[\sqrt{n}]$ so that $n=j^2+m$, $0\leqslant m\leqslant 2j$, and so at j our function is $2j+[m/j]$, while at $j+1$ it is $2j+[(m+1)/(j+1)]$. Both of these [ ]'s are 0 for $m<j$, both are 1 for $j\leqslant m<2j$, but for $m=2j$ the first is 2, while the second is 1. So our minimum is $2j$ for $j^2\leqslant n<j^2+j$ and $2j+1$ for $j^2+j\leqslant n\leqslant j^2+2j$, and this exactly fits the formula $[\sqrt{4n+1}]$.

46. The condition is surely necessary since $1/\alpha$ and $1/\beta$ are the densities of the two sets. In fact, this density consideration really tells the whole story if we look a little closer. The counting function of the first set, i.e., the number of them which are below N, is exactly $[(N+1)/\alpha]$. Similarly, we get $[(N+1)/\beta]$ for the second set. Thus the total number of

terms, either $[n\alpha]$ or $[n\beta]$ which are below N, is $[(N+1)/\alpha]+[(N+1/\beta]$ (counting multiplicity). This is $<(N+1)/\alpha+(N+1)/\beta=N+1$, and is $>(N+1)/\alpha-1+(N+1)/\beta-1=N+1-2=N-1$. So this total number is exactly N. Conclusion: these terms come in *one at a time* and fill up the integers, exactly our sufficiency statement.

47. A moment's thought shows us that what we are asking for is a sequence of positive integers $\{A_n\}$ which is complementary to $\{A_n+n\}$ (and where the A_n are strictly increasing). If we recall Problem 46, we see that the sequences $\{[n\alpha]\}$ and $\{[n\alpha]+n\}$ are complementary provided α is irrational and $1/\alpha+1/(\alpha+1)=1$. This equation leads to the "golden ratio" for α, i.e., $\alpha=(\sqrt{5}+1)/2$, which is irrational. So the formula is $a_n=[n(\sqrt{5}+1)/2]$.

48. This is dazzling! We start with 8,9. Now suppose that $n, n+1$ are both square-full. Then so are $4n(n+1), 4n(n+1)+1$.

Q.E.D.

(That's fine, but how about three consecutive square-fulls? Not even one such is known! And how about consecutive cube-fulls? Always there are more questions than answers!)

49. If we had said *disjoint* rather than *almost disjoint* the answer would be obvious. Each set would be identified by any one of its members and so the collection would be limited to the cardinality of the integers ($\aleph_0$). This way it is problematical.

But, on the other hand, if we had said *rationals* instead of *integers* the problem would become transparent. Every real number has a set of rationals which converge to it, *but* there are c such and any two are obviously almost disjoint! The problem is solved! The rationals are the *same* as the integers, quantitatively.

50. Suppose we begin by letting the first man pour what he thinks is one fair portion, and keep it unless someone else thinks it is too big. If this happens, then each of the others takes a turn pouring back until it looks fair to him. The last one to pour back keeps it. The problem is solved!

The three might just as well have been an n since this method is general. After this first round, namely, we are reduced to the $n-1$ case.

51. The fact that $x_8 = 7$, an integer, is thereby proclaimed an accident! So the sequence is $0, 1, 1, \frac{3}{2}, 2, \frac{11}{4}, \frac{15}{4}, \frac{41}{8}, 7, \ldots$ and the integers are x_0, x_1, x_2, x_4, x_8, the claim being that there are no others. There seems to be a perplexing irregularity to the denominators here, $1, 1, 1, 2, 1, 4, 4, 8, 1, \ldots$? They increase "over-all" but they "pop" down occasionally. A similar pop-down occurs from x_{15} to x_{16}, the denominators being 128 and then 8.

To get a closer look at this sequence we write down the solutions to the recurrence relation—thus if we try ξ^n, we obtain the quadratic $\xi^2 = \xi + \frac{1}{2}$ with the solutions $\xi = (1 \pm \sqrt{3})/2$ and so $A((1+\sqrt{3})/2)^n + B((1-\sqrt{3})/2)^n$ is a solution. With $A = 1/\sqrt{3}$, $B = -1/\sqrt{3}$ the initial conditions, 0 at $n = 0$, 1 at $n = 1$, are satisfied and we conclude that, for all n,

$$x_n = \frac{1}{\sqrt{3}}\left(\frac{1+\sqrt{3}}{2}\right)^n - \frac{1}{\sqrt{3}}\left(\frac{1-\sqrt{3}}{2}\right)^n = \frac{1}{2^{n-1}}\sum_v \binom{n}{2v+1} 3^v.$$

This is the solution we referred to, but it does not seem to help at all to establish the non-integrity. A *breakthrough* occurs if we note that $((1 \pm \sqrt{3})/2)^2 = 1 \pm \sqrt{3}/2$. Thereby we find that

$$x_{2n} = \frac{1}{\sqrt{3}}\left(1 + \frac{\sqrt{3}}{2}\right)^n - \frac{1}{\sqrt{3}}\left(1 - \frac{\sqrt{3}}{2}\right)^n = \binom{n}{1} + \binom{n}{3}\frac{3}{4} + \cdots$$

and this is a formula with a different flavor! The *last* term's denominator has good reason for not being cancelled by the other terms. This is especially obvious for n odd, when this last term is $(\frac{3}{4})^{(n-1)/2}$. Since there is a similar formula for x_{2n+1}

based on $((1\pm\sqrt{3})/2)^{2n+1}=((1\pm\sqrt{3})/2)(1\pm\sqrt{3}/2)^n$, we do feel that the solution is at hand. To facilitate the bookkeeping we break the cases up (mod 4) and write the expansion "backwards." Thus, we have

$$x_{4n}=\binom{2n}{1}\left(\frac{3}{4}\right)^{n-1}+\binom{2n}{3}\left(\frac{3}{4}\right)^{n-2}+\cdots,$$

$$x_{4n+1}=\left(\frac{3}{4}\right)^{n}+\frac{1}{2}\binom{2n}{1}\left(\frac{3}{4}\right)^{n-1}+\binom{2n}{2}\left(\frac{3}{4}\right)^{n-1}$$
$$+\frac{1}{2}\binom{2n}{3}\left(\frac{3}{4}\right)^{n-2}+\cdots,$$

$$x_{4n+2}=\left(\frac{3}{4}\right)^{n}+\binom{2n+1}{2}\left(\frac{3}{4}\right)^{n-1}+\cdots,$$

$$x_{4n+3}=\frac{1}{2}\left(\frac{3}{4}\right)^{n}+\binom{2n+1}{1}\left(\frac{3}{4}\right)^{n}+\frac{1}{2}\binom{2n+1}{2}\left(\frac{3}{4}\right)^{n-1}$$
$$+\binom{2n+1}{3}\left(\frac{3}{4}\right)^{n-1}+\cdots.$$

The last three cases are quite obvious and the "uncancelled" denominators are respectively $2^{2n}, 2^{2n}, 2^{2n+1}$. The first case takes just a bit more care. We must look at the exact power of 2 dividing n. So suppose $2^k|n, 2^{k+1}\nmid n$; the first term then has the denominator 2^{2n-k-3}, and we will show that this too is the net denominator. (This will then suffice since for $n>2$ we have $2n-k-3\geqslant 2n-\log_2 n-3>0$.) We also observe in the presence of k the reason for the observed "pop-down" phenomenon.

So let us observe that any term, past the first one, has the form

$$\binom{2n}{2j+1}\left(\frac{3}{4}\right)^{n-j-1}=\frac{2n}{2j+1}\binom{2n-1}{2j}\left(\frac{3}{4}\right)^{n-j-1},\qquad j>0,$$

so that the denominator, which is a power of 2, can be at most $4^{n-j-1}\cdot 2^{-1-k}=2^{2n-k-3-2j}$. Indeed, then, the leading denominator is *unscathed*.

52. The expression $(1-1/a_1)\cdots(1-1/a_k)$ has a probabilistic ring to it. Indeed each of the factors $1-1/a_i$ is exactly the "probability" that an integer be a non-multiple of a_i. Thus, the result asserts a kind of *super-independence* of these events. If the a_i were pairwise relatively prime, then the events would indeed be independent, and our expression would therefore be exactly equal to this density.

In general we proceed inductively and let S be the set of all positive integers which are not multiples of either $a_1, a_2, \ldots, a_{k-1}$, and we are entitled to assume that S has density $D \geqslant (1-1/a_1)\ldots(1-1/a_{k-1})$. Now let T be the subset of S which are multiples of a_k. We need only show that T has density $\leqslant (1/a_k)\cdot D$ to complete our induction. But consider the set $(1/a_k)\cdot T$ (i.e., comprising all the members of T divided by a_k). Because of the special definition of S (a "sieved" set), $(1/a_k)\cdot T$ is a subset of S. Hence, the density of $(1/a_k)\cdot T$ is $\leqslant D$, and this obviously shows what we needed, that the density of T is $\leqslant D/a_k$.

53. The crucial observation is that if the data "alternates in sign," then the fundamental (Lagrange) interpolating polynomial of degree $n-1$ automatically has all its zeros real since it has $n-1$ sign changes. So, for the general case, all we need do is introduce certain new data points between any two consecutive y_i with the same sign. Our new data pairs number at most $n+(n-1)=2n-1$, so the Lagrange interpolator is of degree at most $2n-2$ and has all its zeros real.

54. What is worth a try is a "power" function, $f(x)=x^\alpha$, since then the functional equation simply becomes $1+2^\alpha+3^\alpha=0$. The x drops out completely and we are left with a mere numerical equation! But x^α is troublesome for negative x, so make it $f(x)=|x|^\alpha$. Also, there is no real solution α to our numerical equation, so we must seek complex α, and our continuity condition requires further that $\operatorname{Re}\alpha>0$.

There are rather neat complex variable proofs of the existence of such solutions α, but in line with our general philosophy, we seek a purely elementary approach. So write $\alpha = a + bi$, and obtain the system

$$2^a \sin(b \log 2) + 3^a \sin(b \log 3) = 0,$$
$$2^a \cos(b \log 2) + 3^a \cos(b \log 3) = -1.$$

Now examine the first equation when $a = 0$. It is $\sin(b \log 2) + \sin(b \log 3) = 0$, which can be satisfied by making $b \log 3 = b \log 2 + \pi$, or $b = \pi/\log \frac{3}{2}$. Notice that for this a, b we have $\cos(b \log 2) > 0$, so that $2^a \cos(b \log 2) + 3^a \cos(b \log 3) > -1$.

At the other extreme, what happens to our first equation as $a \to +\infty$? Clearly we are forced to have $\sin(b \log 3) = 0$, so, e.g., $b = 3\pi/\log 3$. The reason for the 3π rather than 2π or 0, etc., is that, for all b with $\pi/\log \frac{3}{2} < b < 3\pi/\log 3$, the first equation is satisfied by a unique positive a. But for this, a, b (i.e., $+\infty$, $3\pi/\log 3$) we have $2^a \cos(b \log 2) + 3^a \cos(b \log 3) = -\infty < -1$.

Our result, then, is a simple consequence of the intermediate value property.

55. The proofs are routine once the "discoveries" are announced, so we will simply make the announcements.
    1. The set of terminating binary numbers, under addition (mod 1).
    2. The rational functions in x, the subfield being the rational functions in x^2.
    3. Formal series $\sum_{n=1}^{\infty} a_n x^{\alpha_n}$, a_n real, $\alpha_n > 0$, α_n increasing to ∞.

(Multiplication and addition being as usual.)

56. This is a very different kind of problem, and a very perplexing one at first. There is so little to work with! What a surprise that so much follows from this mild, if somewhat bizarre, property.

So, how to begin? Perhaps with the observation that the result would be obvious if we knew that n_k was an *increasing* sequence. For then $n_{k+1} > n_{n_k}$ would ensure $k+1 > n_k$, while just being positive integers (and increasing) would ensure $n_k \geqslant k$. (Together, then, we *would* have that $n_k \equiv k$.)

One bit of this desired monotonicity is obvious. This is the fact that n_1 is, indeed, the smallest term of the sequence. For the simple reason that nothing else could be! Every other term is of the form n_{k+1}, and so has a term, namely, n_{n_k}, smaller than it. In fact this shows that n_1 is strictly smaller than all the other terms.

But this fact about n_1 can be extended. Note, namely, that the new sequence $n_2 - 1, n_3 - 1, n_4 - 1, \ldots$ has *exactly* the same properties as the original one. The terms are positive integers since, again, $n_{k+1} > n_{n_k} \geqslant 1$. Also $n_{(n_{k+1}-1)+1} - 1 < n_{(k+1)+1} - 1$ since this simply says that $n_{n_{k+1}} < n_{k+2}$. Thus, $n_2 - 1$ is the unique smallest of these, which is to say n_2 is smaller than $n_3, n_4, \ldots$.

Repeating this argument does indeed give the full monotonicity property that we required, and the proof is complete.

57. If, for example, the set consisted of the three vertices we would choose the path made of the two legs and then the sum of the squares would *equal* the square of the hypotenuse, but note that the other path, one leg and the hypotenuse, would fail. So it is a challenge; some care must be taken.

Aiming for an induction argument we drop the altitude from the right angle and thereby break the original triangle into two right triangles. But now, as is often the case in induction arguments, we must prove a slightly harder result. Namely, let us require also that our path begins at one end of the hypotenuse and ends at the other. Say we so connect up the points in our two triangles. Note that just joining these two paths will *not* quite serve for the desired path in the original triangle. There is now an extra point included, namely, the right angle vertex. But we can get rid of it! Namely, consider the two segments that are connected to this vertex, remove them, and insert the single segment between the two points thus exposed. We have removed two sides of a triangle with an acute angle between them and replaced them by the third side. The sum of the squares has gone down!

The induction is essentially accomplished except for the nuisance that the number of points may not be sufficiently diminished from the original triangle to the two new ones. This situation is corrected, however, by continuing the construction perhaps a few more times.

58. This is a perfect example of a problem where *knowing* the answer makes it trivial, and where guessing the wrong answer could lead to hours of frustration. The point is that the answer to this question is a surprising NO! So imagine the wasted time in trying to prove the affirmative result.

But knowing that it is NO, we know why immediately. It is because of the different number of at-bats. For example $1/100$ beats $0/1$ for the first half of the season and then it hardly matters what we choose for the second half. The simplest numbers are that $1/1$ beats $1/2$. This leads to a season average of $2/101$ for A and $1/3$ for B, so it is NO. (Of course.)

We may carry this lunacy further. Batter A can even bat better than B *every* single day of the season and still end up behind! On odd days they bat $1/1$ and $2/3$, while on even days they bat $1/4$ and $0/1$. At the end of an even numbered season they end up respectively with 400 and 500. Indeed on every day of the season (except the first) the cumulative average of player two is higher!

59. As usual we view this as two different problems. Find an upper bound, and find a lower bound. If these two surrounding bounds are *close* to each other, then we know we have done a

good job and, indeed, both *are* adequate estimates. Thus each separate bound might be unconvincing while the *pair* can prove itself valid.

An Upper Bound [Remember, we can be wasteful! (if we are lucky later)]: Observe that the single term is bounded by the entire sum (since the a_n are all non-negative, since $e^{e^x} = e^1 e^x e^{x^2/2} \ldots$ is a product of series with non-negative coefficients.) So we have $a_n x^n \leqslant e^{e^x}$, a very wasteful inequality indeed, but perhaps we can cut down on this waste by choosing x wisely. Thus, setting $x = 1$ leads to $a_n \leqslant e^e$, a terrible bound since we know that $a_n \to 0$, in fact that $\sqrt[n]{a_n} \to 0$. A wiser choice would be $x = 2$, since this leads to $a_n \leqslant e^{e^2}/2^n$. But now the calculus intrudes! We wish to do the best we can with $a_n \leqslant e^{e^x}/x^n$, and the very best we can do is to minimize e^{e^x}/x^n over $(0, \infty)$. The derivative equation is $xe^x = n$ and this is annoying, a transcendental equation which simply cannot be solved explicitly. *The end of the line*? No! We simply retreat a little. If we cannot express the very best choice of x, we can at least find an approximation. Remember, we can choose x to be anything we please! A reasonable choice is to make $e^x = n$. (Instead of $xe^x = n$. We may excuse this approximation by noting that x is negligible alongside of e^x—but we do not need to excuse anything. We are the boss!). So $x = \log n$—and we obtain $a_n \leqslant (e/\log n)^n$. End of the first problem.

A Lower Bound: We derived our upper bound by the simple expedient of *ignoring* terms. Curiously enough, this same approach can lead to lower bounds. Previously we viewed our *series* $a_0 + a_1 x + a_2 x^2 + \cdots$ and noted that this was $\geqslant a_n x^n$. Now we view our *function*

$$\begin{aligned} e^{e^x} &= e^{1+x+x^2/2!+\cdots} = e^1 \cdot e^x \cdot e^{x^2/2} \cdot e^{x^3/6} \cdots \\ &= \left(1 + \frac{1}{1!} + \frac{1}{2!} + \cdots\right)\left(1 + \frac{x}{1!} + \cdots\right) \\ &\quad \times \left(1 + \frac{x^2/2}{1!} + \cdots\right) \cdots \end{aligned}$$

and again proceed to ignore terms. Thus, by ignoring everything but $1\cdot(1+x/1!+x^2/2!+\cdots)\cdot 1\cdot 1\cdots$ we find that our coefficient of x^n must exceed $1/n!$. We deduce, in other words, that $a_n \geqslant 1/n!$. But we are in the same admirable spot as before; we are the boss as to which terms to ignore. If this time we ignore all but $1\cdot 1\cdot(1+x^2/2(1!)+x^4/2^2(2!)+\cdots)\cdot 1\cdot 1\cdots$, we would obtain the lower bound of $1/2^n(n!)$ for a_{2n}, so at least for even n we would deduce that $a_n \geqslant 1/2^{n/2}(n/2)!$, an improvement over the previous. For odd n we use instead $1\cdot x(1+x^2/2/1! + \cdots)\cdot 1\cdot 1\cdots$ and a simple straightforward generalization leads us to the following construction: choose k and write $n = kq + r$, $0 \leqslant r < k$. Thus, by ignoring all terms except $1\cdot 1 \cdots (x^r/1!)\cdots((x^k/k!)^q(1/q!)\cdot 1\cdots$ we obtain the lower bound

$$a_n \geqslant \frac{1}{k!^q q!} = \frac{1}{k!^{[n/k]}([n/k])!}.$$

And again we are the boss as to which k to choose. We want to minimize $(k!)^{[n/k]}([n/k])!$ but the calculus is of no use on such functions. However, if we bound it (above) by $(k^k)^{n/k}\cdot(n/k)^{n/k} \leqslant (kn^{1/k})^n$, then the calculus is adequate. Thus, to minimize $kn^{1/k}$ we obtain the derivative equation

$$\frac{d}{dk}\left(\log k + \frac{1}{k}\log n\right) = \frac{1}{k} - \frac{1}{k^2}\log n,$$

or $k = \log n$. Plugging back in gives

$$a_n \geqslant \frac{1}{k^n n^{n/k}} = \frac{1}{\log n^n n^{n/\log n}} = \frac{1}{(e\log n)^n}.$$

End of the second problem.

Now that the "wise-wasteful" estimates are in, we find that we *have* been lucky. The result is that $1/(e\log n)^n < a_n < (e/\log n)^n$ and, aside from the $1/e$ vs. e we see that we have zeroed in on the "correct" order of magnitude.

60. That $x_n \to \infty$ is cheap. Clearly, the x_n form an increasing sequence and so there is a limit L (in the extended reals), and this limit must satisfy $L = L + 1/L$, so $L = \infty$.

Now we grope with the real problem of how fast this approach to ∞ is. A point of view which often helps in sequence problems is that of functions. So regard n as the independent variable and $x = x_n$ as the dependent one. The given recurrence relation can then be construed as saying that a unit increment in the independent variable produces an increment in the dependent one of one over this dependent variable. This situation is reminiscent of differential equations and is analogous to the relation $dx/dn = 1/x$, an equation with solution $x = \sqrt{2n + c}$. A good guess, then, is that x_n is around $\sqrt{2n}$. We do not try to justify the steps leading to this guess, but just let it guide our next steps! Namely, set $y_n = x_n^2 - 2n$ and thereby obtain

$$\begin{aligned} y_{n+1} &= x_{n+1}^2 - 2n - 2 = \left(x_n + \frac{1}{x_n}\right)^2 - 2n - 2 \\ &= x_n^2 + 2 + \frac{1}{x_n^2} - 2n - 2 = x_n^2 - 2n + \frac{1}{x_n^2} \\ &= y_n + \frac{1}{y_n + 2n}. \end{aligned}$$

Hence, $y_n < y_{n+1} < y_n + 1/2n$ which tells us inductively that $0 < y_n < \frac{1}{2}\log n$. And so we have $\sqrt{2n} < x_n < \sqrt{2n + \frac{1}{2}\log n}$, a very good asymptotic formula indeed.

61. Whenever an approximation to a *sum* is called for, the idea of a definite *integral* suggests itself. This statement may, at first glance, seem paradoxical. Should not it be that, whenever an approximation to a definite integral is called for, a certain sum is suggested? Yes, of course, but this is no paradox at all; after all, if A is close to B, then certainly B is close to A.

The point is that $\Sigma_{n=0}^{\infty} f(n)$ should be close to $\int_0^{\infty} f(t)\,dt$. Forgetting about "how close" we get the feeling that $\Sigma_{n=0}^{\infty} x^{n^2}$ is close to $\int_0^{\infty} x^{t^2} dt$ and, luckily, this definite integral is one with a simple, closed form expression! We have

$$\int_0^{\infty} x^{t^2} dt = \int_0^{\infty} e^{-(\log(1/x))t^2} dt = \frac{1}{\sqrt{\log(1/x)}} \int_0^{\infty} e^{-u^2} du$$

$$= \frac{1}{2}\sqrt{\frac{\pi}{\log(1/x)}}\,.$$

As for the comparison between sums and integrals, let us simply recall the picture of the "rectangles under the curve" and the picture of the "rectangles above the curve." From these drawings we deduce that for $f(t)$ any decreasing function, $\Sigma_{n=1}^{\infty} f(n) \leqslant \int_0^{\infty} f(t)\,dt$ (rectangles beneath) and $\Sigma_{n=0}^{\infty} f(n) \geqslant \int_0^{\infty} f(t)\,dt$ (rectangles above).

For our present problem, then (since x^{t^2} is certainly decreasing in t because $x < 1$), the conclusion is

$$\frac{1}{2}\sqrt{\frac{\pi}{\log(1/x)}} \leqslant \sum_{n=0}^{\infty} x^{n^2} \leqslant 1 + \frac{1}{2}\sqrt{\frac{\pi}{\log(1/x)}}\,.$$

These are very good estimates since the 1 is negligible compared to the $\sqrt{\pi/\log(1/x)}$. Indeed we have, asymptotically,

$$\sum_{n=0}^{\infty} x^{n^2} \sim \frac{1}{2}\sqrt{\frac{\pi}{\log(1/x)}} \sim \frac{1}{2}\sqrt{\frac{\pi}{1-x}}$$

as $x \to 1^-$.

62. The uniqueness of the solution, $x(n)$, results from the fact that $x^n + x$ is increasing (in x), and this is exactly the pivot on which we can make our estimations, namely, any a for which $a^n + a > 1$ serves as an upper bound, and any b with $b^n + b < 1$ serves as a lower bound. The art is to find such an a and b which are "close" to one another. For example, 1 always serves

as an a since $1^n+1>1$, and $1-\epsilon$ *eventually* serves as a b since $(1-\epsilon)^n+1-\epsilon<\epsilon+1-\epsilon=1$ for n large. This at least proves the qualitative assertion $x(n)\to 1$. Now for a more serious look. If we try $1-1/n$ for x we find that

$$x^n+x=\left(1-\frac{1}{n}\right)^n+1-\frac{1}{n}\to\frac{1}{e}+1$$

so that $1-1/n$ is eventually an a; similarly for $1-c/n$, c any constant. Thus, the truth is $1-c(n)/n$, where $c(n)\to\infty$. So let us try $c(n)=\log n$. We obtain

$$\left(1-\frac{\log n}{n}\right)^n<e^{(-\log n)/n\cdot n}=\frac{1}{n},$$

while $1-(\log n)/n<1-1/n$ (for $n>e$), so we see that $1-(\log n)/n$ serves as a b $(x(n)>1-(\log n)/n)$. Now it is only a slight shift (literally) which supplies our upper bound. Namely, for $\rho<1$, consider $(1/1+\rho(\log n)/n)^n$; this is

$$\geqslant\left(\frac{1}{e^{\rho(\log n)/n}}\right)^n=\frac{1}{n^\rho},$$

while

$$\frac{1}{1+\rho(\log n)/n}>1-\rho\frac{\log n}{n}.$$

Thus,

$$\left(\frac{1}{1+\rho(\log n)/n}\right)^n+\frac{1}{1+\rho(\log n)/n}>1$$

for large n, so $1/1+\rho(\log n)/n$ serves as an a, a lower bound. Altogether, then, we see we have determined the asymptotics for $1-x(n)$. It is $(\log n)/n$.

63. This is just a colorful way of writing the sequence problem: $x_{n+1}=\sin x_n$, $x_0=\pi/2$, how does $x_n\to 0$? But if we adopt the same method of guess work as in Problem 60, we would obtain the differential equation $dx/dn=\sin(x)-x$, or very crudely

$dx/dn = -x^3/6$, leading to the solution $x = \sqrt{3/(n+c)}$. At any rate we will simply write $y_n = x_n^{-2} - n/3$, or $x_n = 1/\sqrt{n/3 + y_n}$ and derive the formula

$$y_{n+1} = \csc^2 \frac{1}{\sqrt{n/3 + y_n}} - \frac{n+1}{3} = \frac{n}{3} + y_n + \frac{1}{3} + O\left(\frac{1}{n^2}\right) - \frac{n+1}{3}.$$

So $y_{n+1} = y_n + O(1/n^2)$ and y_n is therefore bounded, so our result is $x_n = \sqrt{3/(n+O(1))}$.

64. If we remember a previous lesson, the sum compares well with the integral, and the integral is explicitly expressible. The details are: $1/(2^t + x)$ is decreasing (in t), so

$$\int_0^\infty \frac{dt}{2^t + x} \leqslant \sum_{n=0}^\infty \frac{1}{2^n + x} \leqslant \frac{1}{1+x} + \int_0^\infty \frac{dt}{2^t + x}.$$

Next

$$\frac{d}{dt}\log(1 + 2^{-t}x) = \log 2 \cdot 2^{-t}x \frac{1}{1 + 2^t x} = \frac{-\log 2 \cdot x}{2^t + x},$$

so that

$$\int_0^\infty \frac{dt}{2^t + x} = \frac{-1}{x \log 2}\log(1 + 2^{-t}x)\Big|_0^\infty = \frac{1}{\log 2}\frac{\log(1+x)}{x}.$$

Finally, then,

$$\frac{\log(1+x)}{x\log 2} = \sum_{n=0}^\infty \frac{1}{2^n + x} \leqslant \frac{1}{1+x} + \frac{\log(1+x)}{x \log 2},$$

and surely

$$\sum_{n=0}^\infty \frac{1}{2^n + x} \sim \frac{\log(1+x)}{x\log 2}.$$

65. An upper estimate is really quite obvious. Each triple contains three different couples and all of these couples are required to

be distinct. Therefore, $3\cdot$(Number of these triples) $\leqslant$ Total number of possible couples $= \binom{n}{2}$. The upper bound is, therefore, $\frac{1}{3}\binom{n}{2}$.

Now what is an obvious lower bound? Suppose, namely, we just kept picking triples subject to the restrictions forced by the previous triples. How long could we thereby continue? Clearly, after K such steps, $3K$ couples will have been created and the next triple will have to avoid containing any of them. We have $3K$ couples. How many new triples could these block? Clearly, each one blocks $n-2$ triples and so there is a possibility that we have blocked $3K(n-2)$ triples. We can only be sure of continuation if $3K(n-2) < \binom{n}{3}$, or $K < \frac{1}{9}\binom{n}{2}$.

In summary, then, the lower bound is $\frac{1}{9}\binom{n}{2}$. If we stopped our analysis here we would only have determined that the order of magnitude is n^2, but we would not have determined the asymptotic answer.

There is a construction which does lead to the asymptotic answer, however. This construction is somewhat ad hoc and *slick*, but still, one should not complain about success. The construction: Represent the n objects as the integers $1, 2, \ldots, n$, and take as our collection all triples (a, b, c) for which $a+b+c$ is divisible by n (i.e., is equal to n or $2n$).

Clearly these triples are uniquely determined as soon as two of their members are specified and so they certainly satisfy our requirement. To estimate their number note that there are n choices for a, then there are at least $n-3$ choices for b (the 3 to avoid $a=b$, $b=c$, or $a=c$) and so the number is $\geqslant n\cdot(n-3)\cdot 1/6$. The asymptotic answer is thereby determined as $n^2/6$.

66. As soon as one realizes that it is hopeless to look for a formula (for the nth term of the sequence), it becomes apparent that a more fundamental, less trick-oriented approach is called for. So let us begin with a somewhat unambitious project. The ratio is always bounded by 2 because if N is a member, then $2N$ is and so the element next after N is $\leqslant 2N$. So let us prove that,

eventually, this ratio is bounded by $\frac{3}{2}$. If N is a member, and is even, then $\frac{3}{2}N$ is also a member and we are again done. But what if N is odd? The answer is simply that then we can multiply it by $\frac{4}{3}$ and $\frac{4}{3}N$ will again be a member (as long as $N>1$).

Let us go a step further and show that the ratio is eventually bounded by $\frac{4}{3}$. As we just saw, if $3|N$, then $\frac{4}{3}N$ is a member and so $N'\leqslant\frac{4}{3}N$ and $N'/N\leqslant\frac{4}{3}$. If $3\nmid N$ and N is large enough (at least 8), then $8|N$ and then $\frac{9}{8}N\in S$, $N'\leqslant\frac{9}{8}N$, $N'/N\leqslant\frac{9}{8}<\frac{4}{3}$. If we could continue on in this manner, the problem would be solved. This means that we need only continue the sequence we started 2, $\frac{3}{2}$, $\frac{4}{3}$, $\frac{9}{8}$,... and make sure that it approaches 1. So we proceed by taking advantage of the fact that for any numbers $\alpha>1$ and $K>1$ we can find integers m and n so that $1\leqslant K^m\alpha^{-n}<\sqrt{\alpha}$ (this in turn will allow us to continue; for example, we have $\frac{9}{8}=2^1\cdot(\frac{4}{3})^{-2}$ lies in $(1,\sqrt{\frac{4}{3}}\,)$, etc.) But the requirement may be written as the condition that $[m\xi]$ be even where $\xi=2(\log K)/(\log\alpha)=v+\theta$ ($v=[\xi]$, $0\leqslant\theta<1$).

Now if v is even, we are done, since the choice $m=1$ works. If v is odd, on the other hand, take m as the first integer above $1/(1-\theta)$, and observe that $m\xi=m(v+1)-m(1-\theta)$, so that $[m\xi]=m(v+1)-2$ is indeed even.

67. Another way to look at these unblocked points is that they are the points (m,n), m,n positive integers bounded by N and with gcd equal to 1. Let us call this number $F(N)$. We find we can thereby express the number of these (m,n) with gcd equal to 2, for these points are just the *doubles* of the unblocked points in the square of side $N/2$, and their number is, therefore, $F(N/2)$. Similarly we obtain $F(N/3)$, $F(N/4)$,... for the counts when the gcd's are 3,4,... .

Thus, we obtain $F(N)+F(N/2)+F(N/3)+\cdots=[N]^2$ since the left hand side counts *all* of the lattice points in the original $N\times N$ square. So our work is cut out for us: we must solve the above functional equation for $F(N)$. This is done by the well-known method of Möbius. We have $F(N)+F(N/2)+F(N/3)+F(N/4)+\cdots=[N]^2$ so that $F(N/2)+F(N/4)+\cdots=[N/2]^2$ and subtracting gives $F(N)+F(N/3)+F(N/5)+\cdots=[N]^2-[N/2]^2$.

We next subtract off this functional equation for $N/3$ getting $F(N)+F(N/5)+F(N/7)+\cdots=[N]^2-[N/2]^2-[N/3]+[N/6]^2$. If we imagine this process continued, we see that eventually we will be left with only the term $F(N)$ on the left-hand side and a batch of terms $\pm[N/k]^2$ on the right. We do not try to keep track of this with any great precision (although this could be done) but we simply observe that this right-hand side will be $\sim CN^2$.

Thus, $F(N)\sim CN^2$, and we may revisit the functional equation and obtain

$$CN^2+C\left(\frac{N}{2}\right)^2+C\left(\frac{N}{3}\right)^2+\cdots\sim N^2,$$

$$N^2C\left(1+\frac{1}{2^2}+\frac{1}{3^2}+\cdots\right)\sim N^2,$$

and so

$$C=\frac{1}{1+1/2^2+1/3^2+\cdots}.$$

Euler told us the value of this series, however, and it is $\pi^2/6$. Conclusion: $F(N)\sim(6/\pi^2)N^2$. A charming way of stating this result is that the "probability" of two integers being coprime is $6/\pi^2$.

68. The way of tallying lattice points in a large area is by associating them each with a unit area. We may associate each lattice point, (m,n), with the unit square $(|x-m|\leqslant\frac{1}{2},\ |y-n|\leqslant\frac{1}{2})$ centered there. These unit squares more or less *tile* the given area, so we are led to the approximate count: number of lattice points equals the area. To see how accurate this is we must look near the boundary, where this tiling breaks down.

In the present case all of the associated area lies inside the circle of radius $R+\sqrt{\frac{1}{2}}$, while the associated area certainly contains the disc of radius $R-\sqrt{\frac{1}{2}}$. Thus, we obtain $\pi(R-\sqrt{\frac{1}{2}})^2 \leqslant$ Number of lattice points $\leqslant \pi(R+\sqrt{\frac{1}{2}})^2$. Therefore the required number is $\pi R^2 + O(R)$, and so certainly $\sim \pi R^2$. A rule of thumb is that the lattice point number = Area + O(perimeter).

Second Solution (very similar, but *formally* different): Look at the Riemann sum for the double integral representing the area of the unit disc. That is, we take the function $f(x, y)=1$ for $x^2+y^2 \leqslant 1$, $=0$ for $x^2+y^2>1$, and let the disection points be $(i/R, j/R)$, i, j integers. The mesh size is then $\sqrt{2}/R$ which goes to 0 with $R \to \infty$, so the theory of the Riemann integral tells us that

$$\sum_{i,j} \frac{1}{R^2} f\left(\frac{i}{R}, \frac{j}{R}\right) \to \int\int f(x, y)\,dx\,dy \qquad \text{as } R \to \infty.$$

But the left side is $(1/R^2)\cdot$number of lattice points, and the right side is π. (As before.)

69. As indicated in the instructions we put together two triangles to form a parallelogram whose parallel translations then tile the plane. These parallelograms have lattice point vertices and contain no other lattice point in or on them. If we associate each parallelogram with, say, its southwest corner, then these parallelograms are in 1–1 correspondence with the lattice points.

Now take a huge square and observe that the number of lattice points it contains, its area, and the number of our parallelograms it contains are asymptotically equal. Thus, a square of area A contains asymptotically A parallelograms. Each parallelogram's area is therefore 1. Each triangle's area is therefore $\frac{1}{2}$.

70. The case of S being the even integers, which led to the formula $\Sigma_S x^n/n! = \frac{1}{2}e^x + \frac{1}{2}e^{-x}$, is easily generalized. Consider, for example, the case of S being the multiples of three, and this time look at e^x, $e^{\omega x}$, $e^{\omega^2 x}$, $1, \omega, \omega^2$ being the *cube* roots of 1, i.e., $\omega = e^{2\pi i/3}$. Then

$$\begin{aligned} e^x + e^{\omega x} + e^{\omega^2 x} &= 1 + \frac{x}{1!} + \frac{x^2}{2!} + \cdots + 1 + \frac{\omega x}{1!} + \frac{\omega^2 x^2}{2!} \\ &\quad + \cdots + 1 + \frac{\omega^2 x}{1!} + \frac{\omega^4 x^2}{2!} + \cdots \\ &= 3 + 3 \cdot \frac{x^3}{3!} + 3 \cdot \frac{x^6}{6!} + \cdots, \end{aligned}$$

the other terms cancelling. We thereby obtain the analogous formula

$$\sum_S \frac{x^n}{n!} = \frac{1}{3}e^x + \frac{1}{3}e^{\omega x} + \frac{1}{3}e^{\omega^2 x},$$

and this easily gives the asymptotic formula $\frac{1}{3}e^x$ since, e.g., $|e^{\omega x}| = e^{\mathrm{Re}\,\omega \cdot x} = e^{-x/2}$. The generalization and the asymptotics are obvious. Indeed for $S =$ multiples of D, we obtain $\Sigma_S x^n/n! \sim (1/D)e^x$.

Second Solution: There is a very handy observation which has virtually nothing to do with the exponential series, but applies to any series $S = a_0 + a_1 + a_2 + \cdots$ with decreasing (non-negative) terms. Namely, for any fixed positive integer D, let us write $S_0 = a_0 + a_D + a_{2D} + \cdots$, $S_1 = a_1 + a_{D+1} + \cdots$, $S_k = a_k + a_{D+k} + a_{2D+k} + \cdots$, so that clearly $S_0 \geqslant S_1 \geqslant S_2 \cdots \geqslant S_D = S_0 - a_0$. Hence, all of the S_k are within a_0 of one another, and so are within a_0 of their average, $(1/D)(S_0 + S_1 + \ldots S_{D-1}) = (1/D)\cdot S$. Thus, if a_0 is negligible in some sense, compared to the full sum S, we see that each S_k is practically equal to $(1/D)\cdot S$.

Marvelous, but what has this got to do with $\Sigma x^n/n!$? This is not, usually, a series with decreasing terms! True, but it is a series with *unimodal* terms, namely, the terms are increasing for

n up to $[x]$ and decreasing for terms above $[x]$. So the previous analysis does apply to the two separate pieces of the series, and we deduce that the S_k are all equal to $(1/D)e^x$ to within twice the maximum term, namely, $x^{[x]}/[x]!$. Since this term, by Stirling's formula, is of the order of $e^x/\sqrt{x}$, we see that this is negligible, and we do obtain the asymptotic formula.

Note, however, that the error term, $e^x/\sqrt{x}$, is nowhere near as good as the *miraculously* small error terms obtained by the use of the roots of unity, but this should come as no surprise. The generality of the method necessarily means that it cannot get best possible results.

71. The first (humble) beginning is the observation that the odd numbers fit the bill. Next we see that this is far from maximal; true we cannot add in any doubles of odd numbers, **but** we can throw in all the quadruples. But why stop there? We can simply repeat this construction and obtain all the powers of 4 times odd numbers. This is certainly a maximal set, but is it maximum or even near maximum? We are only asked for an estimate, but perhaps we have achieved an overkill!

Indeed we have. The set we constructed is a maximum. Our first observation that choosing the odds eliminates the doubles can easily be sharpened. For if we consider the odds and their doubles, the most we can take is the odds. Consider all the odds j and the corresponding numbers $2j$. Of each pair j and $2j$ we can choose only one. The same reasoning applies to the $4j$ and $8j$, the $16j$ and $32j$, etc.

Summing up, then, the maximum set is as described and has the count $n-[n/2]+[n/4]-\cdots$ exactly. Note that this is asymptotically equal to $n(1-\frac{1}{2}+\frac{1}{4}-\cdots)=\frac{2}{3}n$.

72. It makes sense that the perfect squares enormously outweigh all the other powers so that $\{X_n\}$ should be "very much like" $\{n^2\}$, i.e., $X_n \sim n^2$. But how to prove such a thing? It is obviously a

bookkeeping problem: how does one *hide* a truly *negligible* quantity? The answer, in this case, is to look at the *other side*, the inverse function. Let us calculate, then, the number of X_n which are below x. The clear implication being that if this turns out $\sim\sqrt{x}$, then it follows that $X_n \sim n^2$. But now life is easy. The exact count of n for which $X_n \leqslant x$ is $[\sqrt{x}]+[\sqrt[3]{x}]-[\sqrt[6]{x}]\cdots$, etc. This is, in turn, equal to $\sqrt{x} + O(\log x)$ terms, each of which is $O(\sqrt[3]{x})$, so the total tally is $\sqrt{x} + O(\sqrt[3]{x}\ \log x)$. The bookkeeping is accomplished!

73. The problem is not about any fixed set of n points, so to speak, but about the *worst*, most spread out, set. So suppose that we look at the lattice work of n points in the unit square. To be precise pick $N=[\sqrt{n}]$ and consider the points $(i/N, j/N)$, $0 \leqslant i \leqslant N, 0 \leqslant j \leqslant N$. By ignoring some of these we can be left with exactly n points and they will all be separated by at least the distance $1/N$. For these points a connecting path would have to be of length at least $(n-1)\cdot 1/N \geqslant (n-1)/\sqrt{n} \geqslant \sqrt{n} - 1$, our lower bound.

To obtain our upper bound we take our cue from this lattice picture. Thus, let us draw the line across the "bottom," from $(0,0)$ to $(1,0)$ then up to $(1,1/N)$, then back to $(0,1/N)$, up to $(0,2/N)$, over to $(1,2/N)$, etc. Thereby we will have path of total length $N+1+N\cdot(1/N)=N+2$.

Of course this path need not pass through any of our given n points—but it does come *close* to all of them. Indeed it comes within $1/2N$ and so if we simply "dart out" to each point and then dart back we will hit each point and we will only have increased our length by $2n\cdot(1/2N)$.

In all, then, our upper bound is given by at most

$$N+2+\frac{n}{N} \leqslant \sqrt{n} + 2 + \frac{n}{\sqrt{n+1} - 1}$$

$$= \sqrt{n} + 2 + \sqrt{n+1} + 1 \leqslant 2\sqrt{n} + 4.$$

The order of magnitude is established at $\sqrt{n}$. (Of course we could look further and go for the asymptotics, but we prefer not to, since the price is a high one in complexity, and nowhere near as neat and simple as this order of magnitude analysis.)

74. A first glance might lead to a very large upper bound since neither the a nor the b need decrease very much in each step. The couple $(k, 2k-1)$ becomes $(k-1, k)$ and the a has shrunk only by 1. Again, the couple $(k, k+1)$ becomes $(1, k)$ and now the b has only shrunk by 1. But the second glance is more fruitful. We look, not at a nor at b, but at $a+b$.

Judging by the two cases above, this quantity seems to decrease by a factor of at most $\frac{2}{3}$. So let us test this hypothesis: namely, we want to know if $a+b-a[b/a] \leqslant \frac{2}{3}(a+b)$, or equivalently, $a+b \leqslant 3a[b/a]$, or $1+b/a \leqslant 3[b/a]$. But this is trivially true since

$$1+\frac{b}{a} \leqslant 2+\left[\frac{b}{a}\right] \leqslant 2\left[\frac{b}{a}\right]+\left[\frac{b}{a}\right]=3\left[\frac{b}{a}\right].$$

This of course produces an upper bound of $c \log n$, and we must now look for the lower bound. To do so let us simply "run" the process backwards. If we ended with $(0,1)$ this could (minimally) have come from $(1,1)$ which then comes from $(1,2)$ and then $(2,3)$, etc. The etc. of course refers to the fact that we are generating the Fibonacci numbers, or pairs of them. But the fact that these numbers grow exponentially means that they reach the size n in only $c' \log n$ steps, and we do have the lower bound.

(Indeed if we were interested in obtaining the asymptotic answer, i.e., making $c = c'$, we would take the combination $a+\rho b$ rather than $a+b$, where $\rho = (\sqrt{5}-1)/2$. We omit the details, however.)

75. The restriction of *no triangles* means that if two points are connected to each other then they are never *both* connected to anything else. But this means that these two points are involved

altogether in at most $n-1$ connections (one, at most, to the $n-2$ other points, and one between them). Repetition of this observation leads us directly to the upper bound of

$$(n-1)+(n-3)+(n-5)+\cdots = n\left[\frac{n}{2}\right]-\left[\frac{n}{2}\right]^2$$
$$= \left[\frac{n}{2}\right]\cdot\left(n-\left[\frac{n}{2}\right]\right).$$

But writing the upper bound in this way suggests the obvious lower bound construction. Simply divide the set in "half," namely, into $[n/2]$ and $n-[n/2]$ elements and connect all of the first half with all of the second half. Total success, then: our lower bound *equals* our upper bound.

76. The point of the problem is that the obvious approach fails. Removal of the brackets introduces n errors each of potential size 1, and so threatens us with an error of order n, the same size as our "principal" term. We must somehow take advantage of the alternating signs. The observation that shows us how to do this is the fact that the terms with denominators from $1+n/2$ to n are all 1 in size, so these terms add up to at most 1. Similarly, the terms from $1+n/3$ to $n/2$ are all 2 in magnitude, so these add up to at most 2. Proceeding in this way shows us that our sum is equal to

$$\sum_{k=1}^{n/A}(-1)^{k-1}\left[\frac{n}{k}\right]+E, \quad |E|\leqslant 1+2+\cdots+(A-1).$$

If we now remove the brackets and extend the series to ∞ we obtain $n\log 2+nO\left(\frac{A}{n}\right)+O\left(\frac{n}{A}\right)+O(A^2)$. Thus, the error can be made $<\epsilon n$ by choosing A large and fixed. If we were more ambitious we could pick A so as to minimize our total error, which is $n/A+A^2$. The choice is $A=\sqrt[3]{n}$, and the resulting formula is $n\log 2+O(n^{2/3})$.

Clearly the same kind of analysis is sufficient for any alternating (and decreasing) series.

77. Even the finiteness of this number is not a triviality so let us look at some low cases to see the reason for it. Suppose $m = n = 1$ which means we are looking at the equation $(ax + b)e^x + cx + d = 0$ and claiming that it has at most 3 real zeros. (That it could have 3 such zeros is clear and in fact any 3 real numbers could be selected as these since that would simply lead to 3 homogeneous linear conditions on the a, b, c, d. Similarly we see where $mn + m + n$ is always possible for the general case.) But to return to the question at hand, why cannot $(ax + b)e^x + cx + d$ have 4 real zeros? The answers come from Rolle's theorem! If a function had 4 zeros its derivative would have 3 zeros, and then its second derivative would have 2 zeros. In our case this second derivative is $(\alpha x + \beta)e^x$, which is not identically 0, so it cannot have two zeros.

The general case is also handled by a deft use of Rolle's theorem. So let $P(x, y)$ be of degree (m, n), $n \neq 0$, and suppose it to have N real zeros. By Rolle's theorem, then, $(d/dx)^{m+1}P(x, e^x)$ has at least $N - m - 1$ real zeros. Furthermore this new expression is equal to $e^x \cdot Q(x, e^x)$, where $Q(x, y)$ has degree $(m, n-1)$ and is not identically 0. The result is, therefore, obtained by induction on n. $Q(x, e^x)$ has at most $m(n-1) + m + n - 1 = mn + n - 1$ real zeros, so we have $N - m - 1 \leqslant mn + n - 1$, or $N \leqslant mn + m + n$, and the induction is complete.

78. The derivative is a wildly discontinuous operator so that the negative first part of the assertion comes as no surprise. For an example just consider $f(x) = x^2 + \sin x^2$ which has derivative $2x + 2x \cos x^2$. If this is not sufficiently convincing, then look at $f(x) = x^2 + \sin e^x$ $(x^2 + \sin e^{e^x})$. This has derivative $2x + e^x \cdot \cos e^x$ $(2x + e^{x+e^x} \cos e^{e^x})$. But now let us take a good hard look at why one cannot differentiate on asymptotic equality (in general), so that we can perhaps see why one can (sometimes).

The asymptotic assumption states that $f(x) = x^2 + x^2\epsilon(x)$, where $\epsilon(\infty) = 0$. Now we want to get at $f'(x)$ which is

$\lim_{h\to 0}((f(x+h)-f(x))/h)$. Why cannot we just write

$$f(x+h)=(x+h)^2+(x+h)^2\epsilon(x+h),$$
$$f(x)=x^2+x^2\epsilon(x),$$

subtract, divide, and get

$$\frac{f(x+h)-f(x)}{h}=2x+h+\frac{(x+h)^2\epsilon(x+h)}{h}-\frac{x^2\epsilon(x)}{h}?$$

We can. So far so good, but these last terms explode if we now try to let $h\to 0$. (There is no reason to assume that there is any cancellation between them.) Convexity allows us to estimate $f'(x)$ without letting $h\to 0$. Indeed convexity says that for any positive h, $(f(x+h)-f(x))/h$ is an upper bound for $f'(x)$, while for every negative h it is a lower bound. Thus, we can see the possibility of estimating $f'(x)$, both above and below, without the danger of having $h\to 0$.

To work out the details, let us insist that $|h|\leqslant x/2$ and that x is big enough so that $|\epsilon(x+h)|\leqslant \epsilon/4$ for all h, $|h|\leqslant x/2$. Thus, we have

$$f'(x)\leqslant 2x+h+\frac{\epsilon}{h}x^2 \qquad \text{for } h>0,$$

$$f'(x)\geqslant 2x+h+\frac{\epsilon}{h}x^2 \qquad \text{for } h<0,$$

and the proof is completed by choosing the first h as $\sqrt{\epsilon}\,x$ and the second one as $-\sqrt{\epsilon}\,x$.

79. A useful tool for dealing with such questions is the following: If $F(x)$ is any differentiable function, then $\lim_{x\to a}F(x)$ exists as long as $\lim_{x\to a,\, x\in S}F(x)$ exists, S being the set of *critical* points, i.e., points where $F'(x)=0$. This is an almost trivial observation since S contains all the relative maxima and minima of $F(x)$. If S has a as a limit point, then these clearly surround all the F values and the result holds. If not, then $F(x)$ is monotone near a and so the limit again must exist.

Let us apply this tool to $\lim_{x\to 0^+}(f(x)/g(x))$. In S we have

$$0=\left(\frac{f(x)}{g(x)}\right)'=\frac{g'(x)}{g(x)}\left(\frac{f'(x)}{g'(x)}-\frac{f(x)}{g(x)}\right)$$

so that, there, $f'(x)/g'(x)=f(x)/g(x)$, and the limit in question does indeed exist through S.

80. This also succumbs to the principle of Problem 79. Clearly, through S we have $f(x)=f(x)+f'(x)$, so $\lim_{x\to\infty}f(x)$ certainly exists. That it equals 0 must now be checked. We may otherwise assume *wlog* that $f(x)\to L$, $0<L\leqslant\infty$, but then from the hypothesis $f'(x)\to -L$, and in turn this forces $f(x)\to-\infty$ which is a contradiction.

Second Solution: Use l'Hôpital's (ordinary) rule: $f(x)+f'(x)=(f(x)e^x)'/(e^x)'\to 0$ and $e^x\to\infty$. Hence, $f(x)=f(x)e^x/e^x\to 0$, and then of course $f'(x)$ also must $\to 0$.

81. We write the hypothesis as $2y'y''/F(x)+2y'y=0$ and integrate both sides from 0 to some $X>0$. Thus, $\int_0^X(2y'y''/F(x))\,dx+y^2(X)-y^2(0)=0$, but the real point is that we can apply the underappreciated third mean value theorem to this integral. This says that if $f(x)$ is decreasing and positive on $[a,b]$, then $\int_a^b f(x)g(x)\,dx=f(a)\int_a^\xi g(x)\,dx$, $\xi\in[a,b]$. Thus, $(1/F(0))(y'(\xi)^2-y'(0)^2)+y^2(X)-y^2(0)=0$, so $y^2(X)\leqslant y^2(0)+y'(0)^2/F(0)$, which expresses the desired boundedness.

The problem is solved. Just for fun though, let us see what results if we applied this same technique without first dividing out by $F(x)$. So now we have $y'^2(X)-y'^2(0)+\int_0^X F(x)2yy'\,dx=0$, and this time the third mean value theorem applied to the interval in reverse. We obtain, namely,

$$y'^2(X)-y'^2(0)+F(X)\int_\xi^X 2yy'\,dx=0,$$

or

$$y'^2(X) - y'^2(0) + F(X)y^2(X) - F(X)y^2(\xi) = 0,$$

so

$$y'^2(X) \leqslant y'(0)^2 + F(X)y^2(\xi).$$

But we already know that y is bounded. Conclusion:

$$y'^2(X) \leqslant A + BF(X) \text{ or } y'(X) = O\left(\sqrt{F(X)}\right),$$

also an interesting result!

82. The right thing to picture is the graph of $f'(x)$, for both $f(x)$ and $f''(x)$ are related to it geometrically in a simple way. Namely, $f(x)$ is the area under this curve and $f''(x)$ is its slope. We are saying, then, that if a curve has a very high ordinate somewhere and a bounded slope, then it will amass a large area. And indeed it obviously will do so right around this high ordinate!

In a sense nothing more need be said. This really constitutes a proof. Qualitatively, yes, but we can still explore the sizes of the functions involved. (Never rush to leave a problem!) So assume $|f(x)| \leqslant a$, $|f''(x)| \leqslant b$, and that $f'(x_0) = c$ (*wlog* $c > 0$). By the bound, b, on the slope, we see that the whole triangle of height c and base $2c/b$ lies beneath the curve. Thus, there is area $\geqslant c^2/b$ under the curve from $x_0 - c/b$ to $x_0 + c/b$. In other words $f(x_0 + c/b) - f(x_0 - c/b) \geqslant c^2/b$ and, from the bound on $f(x)$, this gives $2a \geqslant c^2/b$. We have proved the quantitative theorem: if $|f(x)| \leqslant a$, and $|f''(x)| \leqslant b$, then $|f'(x)| \leqslant \sqrt{2ab}$.

Still we cannot quit. Is this the *right* bound or just the best we can do so far? To check such a thing, and in general to check whether any inequality is "best possible," we go back and see if we can choose parameters so that the " $\leqslant$ " become " $=$ "; if so, then *yes* it *was* a best possible result. (If *no* then back to the drawing board.)

So let us return to the graph of $f'(x)$ and realize that our previous inequalities becomes *equalities* exactly in the case when this graph is made up of line segments of slopes b or $-b$, and with the triangular areas equal to $2a$ or $-2a$. So we are led to the choice of the zigzag function for $f'(x)$. This function (although not quite in C^2) does have the max modulas of $f(x)$, $f'(x)$, $f''(x)$ satisfying $b^2=2ac$, so this is indeed the best possible.

83. To say that $\sqrt{2}$ satisfies this equation is simply to say that the recurrent sequence $x_{n+1}=\sqrt{2}^{x_n}$, $x_0=1$, converges to 2. The fact of convergence follows from monotonicity, and monotonicity is quite general in such problems. This general lesson is simply that a recurrent sequence defined by $x_{n+1}=f(x_n)$ is *always* a monotone sequence if f is an increasing function. The proof is a trivial induction (somehow the fact is not sufficiently well known). Thus, a limit always exists (in the extended reals), and of course must be a fixed point of $f(x)$. In the case of $f(x)=\sqrt{2}^x$ the fixed points are $2, 4, \infty$; for $x_0=1$ the convergence is to 2 as we said. (In general for $x_0<4$ the convergence is to 2, for $x_0=4$ the sequence is constantly equal to 4 and for $x_0>4$ the "convergence" is to ∞.)

Now consider the equation $x^{x^{x^{\cdot^{\cdot^{\cdot}}}}}=4$. Since the exponent of x is again $x^{x^{x^{\cdot^{\cdot^{\cdot}}}}}$ $(\infty-1=\infty)$, we conclude that any solution must satisfy $x^4=4$ so that $x=\sqrt[4]{4}=\sqrt{2}$, but, as we saw, $\sqrt{2}^{\sqrt{2}^{\cdot^{\cdot^{\cdot}}}}=2$, not 4, so there is no solution. The break point is e, and the analysis is very much as before. If $1\leqslant t\leqslant e$, then the sequence is $x_{n+1}=t^{x_n/t}$, $x_0=1$, and this increases and so increases to t. But if $\lambda>e$ then $\lambda^{1/\lambda}<e^{1/e}$ and so $\lambda^{1/\lambda}=t^{1/t}$ for some $t\leqslant e$, so the sequence, $x_{n+1}=\lambda^{x_n/\lambda}$, $x_0=1$ converges to t (not λ!).

84. The whole point of the problem is that we do not assume that $f(x)$ is C^1 or even differentiable. (If we did, it would be trivial.)

Continuity is all we have to work with. If $f(x)$ were *not* constant we would have, *wlog*, $f(a) > f(b)$ for some $a < b$. We then subtract off the linear function to get $g(x)$ with $g(a) = g(b)$. For this $g(x)$ we have

$$\lim_{h \to 0^+} \frac{g(x+2h) - g(x+h)}{h} = \lambda > 0 \qquad \text{for all } x.$$

Our idea now is to apply this fact at *one* special value of x and derive a contradiction.

What special x? Experience tells us to look at the maximum point or the minimum point. So try the maximum, i.e., let x be a point in $[a, b)$ (sic!) where $g(x)$ takes its maximum over $[a, b]$. The aforementioned fact that

$$\lim_{h \to 0^+} \frac{g(x+2h) - g(x+h)}{h} = \lambda > 0$$

tells us that, for small enough h, $g(x+2h) > g(x+h)$, and we can make sure that h is also small enough to ensure $x + 2h < b$. But now we have $g(x+2h) > g(x+h)$ and also, writing $h/2$ for h, $g(x+h) > g(x+h/2)$. Continuing, we have $g(x+h/2) > g(x+h/4), \ldots\ g(x+h/2^n) > g(x+h/2^{n+1}), \ldots$. Again by continuity we have $g(x+h/2^n) \to g(x)$, so we may conclude that $g(x+2h) > g(x)$. This is, indeed, the sought-after contradiction: it says that x was *not* the maximum point.

85. As usual, one half is easy. Namely, if this limit exists uniformly then said limit, $f'(x)$, is the uniform limit of continuous functions, and as such is itself continuous, i.e., $f(x) \in C^1$. To prove the converse, all we need recall is the mean value theorem: $(f(x+h) - f(x))/h = f'(x+\theta h)$, $0 \leqslant \theta \leqslant 1$. By the continuity, and hence uniform continuity, of $f'(x)$, this is within ϵ of $f'(x)$ as soon as h, and so θh, lies in $(-\delta, \delta)$. The uniform convergence to $f'(x)$ is established.

86. Life would be easy if $\sin\theta\cdot\sin 2\theta$ were maximized at $\pi/3$, for then so would $|\sin 4\theta\cdot\sin 8\theta|$, etc., and the problem would be solved. Of course, life is not easy (this function is instead maximized at $\sin^{-1}\sqrt{\frac{2}{3}}$). So we look for the next best thing. What if $\sin\theta\cdot\sin^{\alpha}2\theta$ were maximized at $\pi/3$? Wouldn't that also do the job? *Yes*, it would, as long as $\alpha\in(0,1)$, for we would have $f(\theta)=(\sin\theta\ \sin^{\alpha}2\theta)\cdot(\sin 2\theta\ \sin^{\alpha}4\theta)^{1-\alpha}\cdot(\sin 4\theta\ \sin^{\alpha}8\theta)^{1-\alpha+\alpha^2}\ldots$ (with an oddball term at the end.) Each of the parenthetical factors would be maximized (in modulus) at $\pi/3$, etc. This time we take the *masculine* (?) approach: do not just wish for such an α, *force* it! Thus, we must force

$$\begin{aligned}0&=\frac{d}{d\theta}(\log\sin\theta+\alpha\log\sin 2\theta)\\&=\cot\theta+2\alpha\cot 2\theta,\quad \text{at } \theta=\pi/3.\end{aligned}$$

In other words we want $0=1/\sqrt{3}+2\alpha(-1/\sqrt{3})$, or $\alpha=\frac{1}{2}$. Fine, but now we must go back and fill in the "etc." that was alluded to above. So write $g(\theta)=|\sin\theta\ \sin^{1/2}2\theta|$ and obtain

$$|f(\theta)|=g(\theta)g^{1/2}(2\theta)g^{3/4}(4\theta)\cdots g^{E}(2^{n-1}\theta)\cdot|\sin 2^{n}\theta|^{1-E/2},$$

where $E=\frac{2}{3}(1-(-\frac{1}{2})^{n})$. In particular,

$$\left|f\left(\frac{\pi}{3}\right)\right|=g\left(\frac{\pi}{3}\right)g^{1/2}\left(2\frac{\pi}{3}\right)\ldots g^{E}\left(2^{n-1}\frac{\pi}{3}\right)\left|\sin 2^{n}\frac{\pi}{3}\right|^{1-E/2}.$$

Since g is maximized at $\pi/3$ (and so at $2\pi/3$, $4\pi/3$, $8\pi/3$, etc.) we obtain, by division,

$$\left|\frac{f(\theta)}{f(\pi/3)}\right|\leqslant\left|\frac{\sin 2^{n}\theta}{\sin 2^{n}\pi/3}\right|^{1-E/2}\leqslant\left(\frac{1}{\sqrt{3}/2}\right)^{1-E/2}\leqslant\frac{2}{\sqrt{3}},$$

and this is exactly what was asserted.

87. Repeated application gives $a_2<a_3+a_4<a_4+a_5+a_9+a_{16}<\ldots$, and we denote the resulting sequences of the subscripts as the S_k. So $S_1=(2)$, $S_2=(3,4)$, $S_3=(4,5,9,16),\ldots$. We wish to show that none of these sequences contain a repeated element. Divergence would then be immediate for we could conclude that the "tails" would all exceed a_2.

So suppose otherwise, that a repeat first occurs in S_k. Clearly, this element must be of the form n^2, having come from n and n^2-1 in S_{k-1}. But if $n^2-1 \in S_{k-1}$, then $n^2-2 \in S_{k-2}$, etc., all the way down to $n^2-2n+2 \in S_{k-2n+2}$, since none of the numbers $n^2-1, n^2-2, \dots, n^2-2n+2$ are perfect squares. Thus, $k > 2n-2$, while from $n \in S_{k-1}$ we conclude that $n > k-1$. These two inequalities are clearly contradictory since $n \geqslant 2$.

88. Notice what is not required: we do not ask for these two cubes to be the same as one another, we do not ask for the entries to be bounded by 6, and we do not ask that the entries be distinct from one another on each cube. If we had required any of these, the job would indeed be impossible.

The slickest way to construct these dice is by use of generating functions. Thus, if we think of an ordinary die as the polynomial $x^1+x^2+x^3+x^4+x^5+x^6$, then the outcomes of the ordinary pair of dice show themselves when we multiply $(x^1+x^2+\cdots+x^6)\cdot(x^1+x^2+\cdots+x^6) = 1\cdot x^2+2\cdot x^3+3\cdot x^4+\cdots+6\cdot x^7+5x^8+4x^9+\cdots+1\cdot x^{12}$.

In these terms, then, we require a_i and b_j such that $(x^{a_1}+\cdots+x^{a_6})(x^{b_1}+\cdots+x^{b_6}) = (x+\cdots+x^6)(x+\cdots+x^6)$. Thus, we are seeking an alternate factorization for the right-hand side. Now

$$x+\cdots+x^6 = x\frac{1-x^6}{1-x} = x\frac{1-x^3}{1-x}(1+x^3)$$
$$= x(1+x+x^2)(1+x)(1-x+x^2).$$

The pair of factors we desire must multiply, then, to $x^2(1+x+x)^2(1+x)^2(1-x+x^2)^2$. Each must have an x (the condition of having *positive* entries) and each must have a $1+x$ and $1+x+x^2$ (the condition of having 6 faces, meaning that the value at $x=1$ is 6). The only free choice is how to distribute the two $1-x+x^2$ factors. If we give one to each we get ordinary dice so we must try the other way. Thus, we take

$$x(1+x+x^2)(1+x) = x+2x^2+2x^3+x^4,$$

$$x(1+x+x^2)(1+x)(1-x+x^2)^2 = x+x^3+x^4+x^5+x^6+x^8.$$

This is the solution. One die reads $1,2,2,3,3,4$; the other reads $1,3,4,5,6,8$.

89. We begin by experimenting. Put $0 \in A$, and then everything seems determined. Thus $1 \in B$, or else 1 would be expressible by $a+a'$ but not by $b+b'$. Next, $2 \in B$, or else 2 would be $a+a'$ but not $b+b'$. Then $3 \in A$, since $3 = b+b' = 1+2$, etc. The sets therefore start out looking like: $A = \{0,3,5,6,9,10,12,15,\ldots\}$, $B = \{1,2,4,7,8,11,13,14,\ldots\}$. (Of course there is no proof that this partition is possible, or, if possible, that it is unique.) It might be possible to spot the rule of formation for A and B and then, equipped with this right guess, prove all by induction. However, we elect to not look for such a brilliancy but to rely on our method of generating functions. So let our hypothetical generating functions be $A(x) = \Sigma_{a \in A} x^a$, $B(x) = \Sigma_{b \in B} x^b$, and then observe that our defining conditions become $A(x) + B(x) = 1/(1-x)$ and $A^2(x) - A(x^2) = B^2(x) - B(x^2)$. Thus,

$$(A(x) - B(x))(A(x) + B(x)) = A(x^2) - B(x^2),$$

so

$$A(x) - B(x) = (1-x)(A(x^2) - B(x^2)).$$

Repeating gives

$$A(x) - B(x) = \prod_{n=0}^{N-1} (1 - x^{2^n})(A(x^{2^N}) - B(x^{2^N})),$$

and letting $N \to \infty$ gives

$$A(x) - B(x) = \prod_{n-0}^{\infty} (1 - x^{2^n}).$$

But this product, when multiplied out, gives $+x^k$ if k is the sum of an even number of distinct powers of 2 and $-x^k$ if k is an odd number of them.

Of course this means that

$A =$ Set of all integers with an even number of 1 digits in its binary representation.

$B =$ Those with an odd number of 1 digits.

Indeed this fits with our experimental evidence

$$A = \{0,3,5,6,9,10,12,15,\ldots\},$$
$$B = \{1,2,4,7,8,11,13,14,\ldots\},$$

and this time we have proven the existence and uniqueness.

90. The generating function for the arithmetic progression $an + b$ is the series $\Sigma x^{an+b} = x^b/(1-x^a)$. What we are asking is whether we can have $x/(1-x) = \Sigma_{j=1}^{k}(x^{bj}/(1-x^aj)$ with $k \geqslant 2$ and all a_j distinct. The answer is *no*, as we see by setting $x = e^{2\pi i/a}$ where $a = \text{Max}_{1 \leqslant j \leqslant k}(a_j)$ (or rather by letting $x \to e^{2\pi i/a}$ since we really must keep $|x| < 1$ in order to have a bona fide series.) The point is that at $e^{2\pi i/a}$ one of these terms is ∞ and all of the others are finite, a genuine contradiction.

91. No. If the hypothetical set is A then our generating function is $\Sigma_{a \in A} x^a$, and our condition is simply that the series for $((\Sigma x^a)^2 + \Sigma x^{2a})/2$ have all its coefficients equal after awhile. This is to say that $(\Sigma x^a)^2 + \Sigma x^{2a} = P(x) + C/(1-x)$ with $P(x)$ a polynomial.

A contradiction is obtained by letting $x \to (-1)^+$ i.e., along the negative axis. The left-hand terms are non-negative and approaching ∞, while the right-hand side remains bounded.

92. *No*. Suppose that the numbers in question are $a_1, a_2, \ldots, a_n$. The number of positive differences is then $n(n-1)/2$. Thus, we are requiring them to be the integers $1,2,3,\ldots,n(n-1)/2$ each taken on once. If we form the generating function $x^{a_1} + x^{a_2}$

$+\cdots+x^{a_n}$, then the differences are given by the exponents in the product $(x^{a_1}+\cdots+x^{a_n})(x^{-a_1}+\cdots+x^{-a_n})$. We are requiring that all the terms $x+x^2+\cdots+x^{n(n-1)/2}$ appear precisely once. But the term 1, i.e., x^0, must appear n times and all of the "negative" terms $x^{-1}+x^{-2}+\cdots+x^{-n(n-1)/2}$ must also appear. In short, our hypothetical requirement is exactly that

$$\begin{aligned}(x^{a_1}+\cdots+x^{a_n})&(x^{-a_1}+\cdots+x^{-a_n})\\ &=x^{-n(n-1)/2}+\cdots+x^{-1}+n+x+x^2+\cdots+x^{n(n-1)/2}\\ &=n-1+\frac{x^{-n(n-1)/2}-x^{n(n-1)/2+1}}{1-x}.\end{aligned}$$

The search for a contradiction is not so easy this time. Nothing ever goes to ∞! True, we can set x equal to anything we please but where should we look? The hint comes from examining the left-hand side and noticing that when x is on the unit circle, i.e., $x=e^{i\theta}$, the two factors become complex conjugates. So let us try this substitution and observe that then

$$\frac{x^{-n(n-1)/2}-x^{n(n-1)/2+1}}{1-x}=\frac{\sin((n^2-n+1)/2)\theta}{\sin(\theta/2)},$$

and our whole requirement translates into

$$|e^{ia_1\theta}+\cdots+e^{ia_n\theta}|^2=n-1+\frac{\sin((n^2-n+1)/2)\theta}{\sin(\theta/2)}.$$

But what is the contradiction? This was seen to be possible for $n=4$, i.e., we do have

$$|1+e^{2i\theta}+e^{5i\theta}+e^{6i\theta}|^2=3+\frac{\sin(13/2)\theta}{\sin(\theta/2)}.$$

But for larger n perhaps we can make the term

$$\frac{\sin((n^2-n+1)/2)\theta}{\sin(\theta/2)}$$

very large negative. Indeed a good choice would be to make the numerator equal -1. So choose $\theta = 3\pi/(n^2 - n + 1)$; this makes the numerator -1 and the denominator $< 3\pi/2(n^2 - n + 1)$. Then we obtain $0 \leqslant n - 1 - 2(n^2 - n + 1)/3\pi$, and this is surely a contradiction for large n, but in fact it is a contradiction from $n = 5$ on, for

$$n - 1 - \frac{2(n^2 - n + 1)}{3\pi} < n - 1 - \frac{2(n^2 - n)}{3\pi} = (n-1)\left(1 - \frac{2n}{3\pi}\right)$$
$$\leqslant (n-1)\left(1 - \frac{10}{3\pi}\right) < 0.$$

93. We write the two generating functions and prove them equal. First do the partitions into odds. Here the generating function is

$$(1 + x + x^2 + x^3 + \cdots)(1 + x^3 + x^6 + \cdots)(1 + x^5 + x^{10} + \cdots)$$
$$\cdots (1 + x^{2m+1} + x^{2(2m+1)} + \cdots) \cdots$$

(for example the partition of $15 = 1 + 1 + 3 + 5 + 5$ corresponds to the term in the product of $x^2 \cdot x^3 \cdot x^{10}$, etc.) which simplifies to

$$\frac{1}{1-x} \cdot \frac{1}{1-x^3} \cdot \frac{1}{1-x^5} \cdot \cdots .$$

Now do the partitions into distinct numbers. This time we get instead $(1+x)(1+x^2)(1+x^3)\cdots(1+x^n)\cdots$, so our problem reduces to proving that these two infinite products are equal.

Put another way, we must prove that

$$(1-x)(1-x^3)(1-x^5)\cdots(1+x)(1+x^2)(1+x^3)\cdots \equiv 1.$$

To do so, observe what happens if we take the factors $(1+x)$, $(1+x^3), (1+x^5)\ldots$ and multiply them respectively by the factors $(1-x), (1-x^3), (1-x^5), \ldots$. What happens is that our product becomes

$$(1-x^2)(1-x^6)(1-x^{10})\cdots(1+x^2)(1+x^4)(1+x^6)\cdots,$$

the very same product as we had before, but with x^2 replacing x!

So, calling the product $P(x)$, we see that $P(x) = P(x^2)$. Repeating this relation we get $P(x) = P(x^{2^n})$, and letting $n \to \infty$ (for $|x| < 1$) we obtain $P(x) = P(0) = 1$, which is exactly what we wanted to prove.

94. The integrand is approaching $e^x \cdot e^{-x} = 1$ at all points and so (formally) the integral approaches $\int_0^\infty 1\, dx = \infty$. That this approach is actual and not just formal is also simple to prove (Fatou's lemma), but we do not dwell on this. Rather let us look with more precision at this approach to 1. Namely, we have

$$\log\left\{\left(1+\frac{x}{n}\right)^n e^{-x}\right\} = n \log\left(1+\frac{x}{n}\right) - x$$
$$= n\left(\frac{x}{n} - \frac{1}{2}\frac{x^2}{n^2} + \frac{1}{3}\frac{x^3}{n^3} - \cdots\right) - x$$

for each x, as soon as $n > x$. Hence,

$$\left(1+\frac{x}{n}\right)^n e^{-x} = e^{-(x^2/n)/2+(x^3/n^2)/3-\cdots}$$

and this suggests the substitution of $x = \sqrt{n}\, t$. So, nothing has been proved, nothing has been asserted, but a *move* has been suggested, and we make it. Our integral becomes $\sqrt{n}\int_0^\infty (1 + t/\sqrt{n})^n e^{-\sqrt{n}\,t} dt$, and, as we have seen,

$$\left(1+\frac{t}{\sqrt{n}}\right)^n e^{-\sqrt{n}\,t} = e^{-t^2/2+(t^3/\sqrt{n})/3-(t^4/n)4+\cdots}$$

so that we are tempted to proclaim $\sqrt{n}\int_0^\infty e^{-t^2/2}\, dt = \sqrt{\pi n/2}$ as the asymptotic formula. How can we justify this? The fundamental theorem telling when $\lim_n \int f_n = \int \lim_n f_n$ is the dominated convergence theorem. This says that all that is required is $\int \operatorname{Max}_n |f_n| < \infty$. Thus, although it is not often looked at in this way, we are faced with a calculus-type problem, to maximize $|f_n|$ over n.

In the case at hand, differentiation shows that $(1 + t/\sqrt{n})^n e^{-\sqrt{n t}}$ is decreasing in n (see Note) and so the maximum occurs for $n = 1$ *always*. The maximum is $(1 + t)e^{-t}$, and this is integrable, so reversing of the limit and integral is justified, and our answer is $\int_0^\infty (1+x/n)^n e^{-x}\,dx \sim \sqrt{\pi n/2}$.

(Note: The logarithmic derivative, with respect to n, is

$$\log\left(1+\frac{t}{\sqrt{n}}\right)-\frac{1}{2}\frac{t}{\sqrt{n}+t}-\frac{1}{2}\frac{t}{\sqrt{n}},$$

and if we change variables by setting $\log(1+t/\sqrt{n})=u$, this in turn becomes $u-\sinh u$ which is clearly negative, since u is positive.)

95. Problem 94 shows how to "handle" $\int_0^\infty (1+t/\sqrt{n})^n e^{-\sqrt{n}t}\,dt$. Our present integral gets reduced to something like this one by the change of variables $x=n+\sqrt{n}t$, namely,

$$\int_0^\infty x^n e^{-x}\,dx=\left(\frac{n}{e}\right)^n\sqrt{n}\int_{-\sqrt{n}}^\infty\left(1+\frac{t}{\sqrt{n}}\right)^n e^{-\sqrt{n}t}\,dt.$$

Since we still have

$$\lim_n\left(1+\frac{t}{\sqrt{n}}\right)^n e^{-\sqrt{n}t}=e^{-t^2/2}$$

we do see that, formally,

$$\int_{-\sqrt{n}}^\infty\left(1+\frac{t}{\sqrt{n}}\right)^n e^{-\sqrt{n}t}\,dt\to\int_{-\infty}^\infty e^{-t^2/2}\,dt=\sqrt{2\pi},$$

and Stirling's formula follows. We need to invoke the calculus now to estimate $\mathrm{Sup}_n(1+t/\sqrt{n})^n e^{-\sqrt{n}t}$, where t is any real number and n ranges over all positive integers with $\sqrt{n}\geqslant -t$. As before, the derivative with respect to n tells us that the Sup_n occurs for $n=1$ when $t\geqslant 0$, but another look shows the derivative to be positive for $t<0$, which shows that in that case the Sup occurs as $n\to\infty$, and so has the value $e^{-t^2/2}$. This is also integrable, so dominated convergence holds, and Stirling's formula is established.

96. There is a strong temptation here to apply Stirling's (asymptotic) formula to the factorials involved in this sum, but these lead to cumbersome, annoying expressions, so the temptation is best resisted! There is a much more direct, and hence cleaner, way to begin the assault. The specified sum is, after all, a partial sum of the power series for e^x at $x=n$ and there are definite integral formulas for this. We will end up, then, exchanging a sum for an integral, and this is usually a good bargain.

The formula referred to, of course, is

$$R_n(x)=\int_0^x \frac{(x-t)^n}{n!} f^{(n+1)}(t)\,dt,$$

$R_n(x)$ being the remainder of the Taylor series for $f(x)$. Equivalently, this can be written as

$$s_n(x)=f(x)-\int_0^x \frac{(x-t)^n}{n!} f^{(n+1)}(t) \to dt,$$

$s_n(x)$ being the so-called nth partial sum (meaning the sum of the first $n+1$ terms) of the Taylor series for $f(x)$. Applying this to the function e^x gives $1+x/1!+\cdots+x^n/n!=e^x-\int_0^x(x-t)^n/n!e^t dt$, and setting $x=n$ gives a neat formula for the sum we are considering, viz.,

$$1+\frac{n}{1!}+\cdots+\frac{n^n}{n!}=e^n-\int_0^n \frac{(n-t)^n}{n!} e^t\,dt.$$

We can even neaten this up a bit by changing variables by $n-t=x$ so that the right-hand side becomes $e^n-e^n\int_0^n(x^n/n!)e^{-x}\,dx$, and, ever more neatness, this is equal to $(e^n/n!)\int_n^\infty x^n e^{-x}\,dx$, since $\int_0^\infty x^n/n!e^{-x}\,dx=1$ (Euler's Γ-function). The quickest path from here is via a previous problem (93), which told us that $\int_0^\infty(1+x/n)^n e^{-x}\,dx \sim \sqrt{\pi n/2}$, or $\int_0^\infty(n+x)^n e^{-x}\,dx \sim n^n\sqrt{\pi n/2}$, or $\int_n^\infty x^n e^{-x}\,dx \sim (n/e)^n\sqrt{\pi n/2}$, so what we are looking for, $e^n/n!\int_n^\infty x^n e^{-x}\,dx$, is asymptotic to $(e^n/n!)(n/e)^n\sqrt{\pi n/2}$, and Stirling finishes the job. It says that $n!\sim(n/e)^n\sqrt{2\pi n}$, and thereby gives us $(e^n/(n/e)^n\sqrt{2\pi n})\cdot(n/e)^n\sqrt{\pi n/2}=\frac{1}{2}e^n$.

97. As before we note that $\cos x^2 = 1 - x^4/2 + \cdots$, so that

$$\cos^n x^2 = \left(1 - \frac{x^4}{2} + \cdots\right)^n = e^{-nx^4/2} + \text{higher terms.}$$

Set $x = y/\sqrt[4]{n}$, our integral becomes $(1/\sqrt[4]{n})\int_0^{\sqrt[4]{n}} \cos^n(y^2/\sqrt{n})\,dy$, and $\lim_n \cos^n y^2/\sqrt{n} = e^{y^4/2}$, so that dominated convergence would give us the asymptotic answer $(1/\sqrt[4]{n})\int_0^\infty e^{-y^4/2}\,dy$.

To obtain the "dominator" we seek to estimate $\text{Max}_n \cos^n(y^2/\sqrt{n})$, and if we in fact allow n to vary through all positives this is equal to $(\text{Max}_{0<t\leqslant 1} \cos^{1/t^2} t)^{y^4}$ (with $t = y^2/\sqrt{n}$), and we are done since this Max is strictly less than 1. (The only trouble might be as $t \to 0$, but then the limit is clearly $e^{-1/2}$.)

98. Seeking a recurrence relation we denote by $E(m,n)$ this expected length. What happens, then, after one toss? Half the time you win and then our parameters become $m+1$, $n-1$, and the other half the time they become $m-1$, $n+1$. Thus $E(m,n) = 1 + \frac{1}{2}E(m-1,n+1) + \frac{1}{2}E(m+1,n-1)$, the 1 indicating the first toss. Of course this assumes $m, n > 0$. The boundary conditions are that $E(0,n) = 0$ and $E(m,0) = 0$, and these, together with the recurrence formula, do determine uniquely the function $E(m,n)$.

Indeed, if we view $E(m,n)$ as a function of one variable, say n, along the line $m+n =$ constant, then the formula says that the second difference is a constant (-2), and so $E(m,n)$ is a quadratic function. Vanishing at the end points forces this to be $cm \cdot n$, and direct evaluation shows $c = 1$. Our answer then is $m \cdot n$.

99. This is a mathematical expectation problem. Of course, there is a chance that you will never be behind, and also there is a chance that you will be behind by \$1,000,000 at some point. We desire the *average* over all these possibilities. So we set $p_n =$

probability that you are at some time behind by at least \$$n$. We seek a recurrence relation between these p_n, so that we can determine them, the point being that the desired expectation is equal to $\Sigma_{n=0}^{\infty} p_n$.

Consider p_n and view the first toss. If this is a win for you then, from there one, you will have to at some time be behind by \$$(n+1)$. If this first toss is a loss, then, you will afterward have to be behind by at least \$$(n-1)$. Thus, we have the desired recurrence formula: $p_n = 0.51 p_{n+1} + 0.49 p_{n-1}$. But this is a different equation of degree 2, so that to solve it we need two boundary conditions. One obvious condition, of course, is $p_0 = 1$, but what is the other? We do not know p_1 or p_2, etc., but we do know p_∞. Indeed, we have $p_\infty = 0$, so we can solve the difference equation. By induction, we obtain $p_n = a(49/51)^n + 1 - a$, a some constant, and $p_\infty = 0$ forces $a = 1$. Thus, $p_n = (49/51)^n$, so the expectation $= \Sigma_{n=0}^{\infty}(49/51)^n = 51/2$. That is, the answer is \$25.50.

100. This is another application of the failure probabilities formula. If f_n = probability that n such spins have a sum less than 1, our expected number is $\Sigma_{n=0}^{\infty} f_n$. We must now find these f_n, and we do so by writing them as multiple integrals. Namely,

$$f_n = \underset{\substack{x_1 + \cdots + x_n < 1 \\ x_i > 0}}{\int \cdots\cdots \int} dx_1\, dx_2 \ldots dx_n .$$

There are many ways of evaluating this integral, for example change variables by setting $x_n = 1 - t$ so that it becomes

$$\underset{\substack{x_1 + \cdots + x_{n-1} < t \\ x_i > 0}}{\int \cdots\cdots\cdots \int} dx_1 \ldots dx_{n-1}\, dt,$$

and now change these variables by setting $x_i = tx_i$. The result is then

$$\int_0^1 t^{n-1}\left(\int_{\substack{x_1+\cdots+x_{n-1}<1\\ n>0}}\!\!\cdots\!\!\int dx_1\ldots dx_{n-1}\right)dt$$

$$=\frac{1}{n}\int_{\substack{x_1+\cdots+x_{n-1}<1\\ x_i>0}}\!\!\cdots\!\!\int dx_1\ldots dx_{n-1}=\frac{1}{n}f_{n-1},$$

and since $f_0=1$ this recurrence tells us that $f_n=1/n!$.

We therefore obtain the rather charming result that the expected number of spins is e.

Second Solution: Call $E(x)$ the expected number of spins for the sum to exceed x. Since the first spin gives a t which is uniformly distributed on $[0,1]$ we have the relation $E(x)=1+\int_0^1 E(x-t)\,dt$. But $E=0$ for $x<0$, so if $0\leqslant x\leqslant 1$ this relation can be replaced by $E(x)=1+\int_0^x E(x-t)\,dt=1+\int_0^x E(u)\,du$. Hence $E'(x)=E(x)$, $E(0)=1$, and so of course $E(x)=e^x$. In particular, we again obtain $E(1)=e$.

101. We again use the failure probability formula. Note that the probability that k independent choices lead to different outcomes is

$$1\cdot\frac{n-1}{n}\cdot\frac{n-2}{n}\cdots\frac{n-k+1}{n}=\binom{n}{k}\frac{k!}{n^k}.$$

Hence our desired expectation is equal to $\Sigma_k\binom{n}{k}k!/n^k$.

As for the asymptotics, we write this formula as a definite integral by using Euler's Γ-integral, $k!=\int_0^\infty x^k e^{-x}\,dx$ so that

$$\Sigma\binom{n}{k}\frac{k!}{n^k}=\int_0^\infty\left(1+\frac{x}{n}\right)^n e^{-x}\,dx.$$

What luck! This is an integral which we have already evaluated asymptotically (Problem 94), and the answer is $\sqrt{\pi n/2}$. For $n=365$, which is the situation for ordinary birthdays, $\sqrt{(\pi/2)365}=23.9\ldots$. One need only stop 24 people on the average to find 2 with the same birthday.

102. The probability of success after k days is $1-1/2^k$ for each flower, and so for the total garden is $(1-1/2^k)^n$. Thus, the failure probability is $1-(1-1/2^k)^n$, and our formula gives $\sum_{k=0}^{\infty} 1-(1-1/2^k)^n$ for the desired expectation.

As for the asymptotics, we can proceed *automatically* or *thoughtfully*. The automatic approach is the replacement of this sum by the integral, $\int_0^\infty (1-(1-1/2^k)^n)\,dk$, which, by the change of variables $x=1-1/2^k$, is equal to

$$\frac{1}{\log 2}\int_0 (1-x^n)\frac{dx}{1-x}=\frac{1}{\log 2}\left(1+\frac{1}{2}+\frac{1}{3}+\cdots+\frac{1}{n}\right).$$

Hence our sum is $(\log n)/(\log 2)+O(1)$.

As promised we can get this result by direct "common sense." Just observe that the early terms are close to 1 and the later ones are close to 0. Thus, for the early terms we use $1-(1-1/2^k)^n>1-e^{-n/2^k}$, while for the late ones, we use $1-(1-1/2^k)^n<1-(1-n/2^k)=n/2^k$. Thus, if we break up the series at $k=[(\log n)/(\log 2)]$, we obtain

$$\frac{\log n}{\log^2 2}+n\left(\frac{2}{n}+\frac{2}{2n}+\frac{2}{4n}+\cdots\right)=\frac{\log n}{\log 2}+2$$

as an upper bound, and we also obtain

$$\frac{\log n}{\log 2}-1-\sum_{2^k\leqslant n} e^{-n/2^k}>\frac{\log n}{\log 2}-1-\left(e^{-1}+e^{-2}+e^{-4}+\cdots\right)$$

$$>\frac{\log n}{\log 2}-2$$

as the lower one.

103. Since our solution of Problem 8 gave a method with expected number 2, we see that we have a best possible result. We will use our failure probability formula, this time on the hypothetical experiment with success probability $\frac{1}{3}$. The point is that $\frac{1}{3}$ is not a fraction with denominator 2^k, so there cannot be a *guaranteed* success in any finite number of tosses. Thus, each

failure probability is positive. But the kth failure probability is therefore $\geqslant 1/2^k$, so the expectation is $\geqslant \Sigma_0^\infty 1/2^k = 2$, as required.

Note that exactly the same analysis applies to any probability p which is not a binary rational. Combined with Problem 8, then, we see that 2 is again the best possible expectation.

104. The answer is 1. The expected number of correct letters in *each envelope* is $1/n$. Even though these events are not independent, the expectation *is* linear, so our answer is the sum of n numbers, each of which is $1/n$, and so, indeed, is equal to 1.

105. Again we use the heuristic, ϵ, δ free, language, with the clear understanding that the details could (and therefore need not) be supplied. So we are looking at the numbers $a \cdot b$, $a, b \in [1, n]$. By our FACT about prime factors we conclude that "most" numbers, a, have "around" log log a prime factors. Furthermore, for "most" a, log log a is "around" log log n (even log log $a <$ (log log n)-1 means that $a < n^{1/e}$, which is a very small set indeed).

Since the same statement can be made about "most" b, we deduce that most of our $a \cdot b$ have around 2 log log n prime factors. The conclusion, again according to our FACT, is that most of the numbers appearing on the multiplication table are "rare" (they have nearly 2 log log n prime factors). Believe it or not, the proof is complete.

What a strange proof! We find ourselves in possession of a kind of *empty* knowledge. That is, we know that the set of products $a \cdot b$, $a, b \in [1, n]$ has cardinality $o(n^2)$, but we do not know what the set looks like or even what its cardinality really amounts to! Of course this is the price we must pay for the non-quantitative statement that we pivoted on. If we had used some sharper quantitative statements of the FACT, we would have obtained, for example, $O(n^2/\log\log n)$ rather than $o(n^2)$. But how good is this? Again, it is the upper bound versus lower bound game.

The answer is that it is fairly good, for the count is at least $> c(n^2/\log^2 n)$ this being the number of $a \cdot b$ when a and b are primes. (It is even possible to obtain $c(n^2/\log n)$ by requiring only the b to be prime, while $a < b$.)

106. First we give a "formal," or "not too rigorous" or, in short, *wrong* proof. We pick $h(x) = \int_0^1 f(x+a)\,da$ and obtain

$$h'(x) = \frac{d}{dx}\int_x^{x+1} f(t)\,dt = f(x+1) - f(x),$$

which does approach 0 as $x \to \infty$. Also $g(x) = f(x) - h(x) = \int_0^1 (f(x) - f(x+a))\,da$ and, as we predicted, this *formally* goes to 0 with x, since the integrand does go to 0 pointwise. At this point we could "sneak" away with it by going back and adding some bit of hypothesis, such as boundedness of $f(x)$, which would then justify this passage to the limit inside the integral.

For a *non-sneak* solution let us apply our category argument. The sets, of a, for which $|f(x+a) - f(x)| \leqslant 1$ for all $x \geqslant N$ are closed sets and, by our hypothesis fill the line. Thus, we conclude that one of these sets contains an internal, $[\alpha, \beta]$, with $\alpha < \beta$. Now we simply re-do our proof by making $h(x) = (1/(\beta - \alpha)) \int_\alpha^\beta f(x+a)\,da$. Again

$$h'(x) = \frac{f(x+\beta) - f(x+\alpha)}{\beta - \alpha} = \left(\frac{f(x+\beta) - f(x)}{\beta - \alpha}\right) - \left(\frac{f(x+\alpha) - f(x)}{\beta - \alpha}\right) \to 0 \qquad \text{as } x \to \infty,$$

and, this time,

$$f(x) - h(x) = \frac{1}{\beta - \alpha}\int_\alpha^\beta (f(x) - f(x+a))\,da,$$

and passage to the limit *is justified* because we arranged to make the integrand bounded!

Note also (how nice) that the condition is necessary *and* sufficient. If $g(x) \to 0$, then surely $g(x+a) - g(x) \to 0$ for all a, and if $h'(x) \to 0$, then $h(x+a) - h(x) = ah'(x + a\theta) \to 0$ for all a.

107. Yes. Pick $\epsilon > 0$ and look at the set of a such that $|f(na)| \leqslant \epsilon$ for all $n \geqslant N$. These are closed sets and, by our hypothesis, they fill the whole line, as we let N take all values. This is a perfect setup for our "category argument"! Indeed, it follows that, for some N, the set contains an interval, say $[\alpha, \beta]$, $\alpha < \beta$. By the definition of this set, then, we have $|f(x)| \leqslant \epsilon$ for all $x \in [N\alpha, N\beta]$, for all $x \in [(N+1)\alpha, (N+1)\beta]$, and for all $x \in [(N+2)\alpha, (N+2)\beta]$, etc. But as soon as $(N+k)\beta > (N+k+1)\alpha$ these intervals all overlap, so for $x \in [(N+k)\alpha, \infty)$ we have $|f(x)| \leqslant \epsilon$, and isn't this the definition of going to 0?

108. If $f(x)$ is c.m., then so is $f(x+a)$ for any $a > 0$. But perhaps slightly less evident is the fact that so is $f(x) - f(x+a)$. Namely,

$$\left(\frac{-d}{dx}\right)^k (f(x) - f(x+a)) = a\left(\frac{-d}{dx}\right)^{k+1} f \quad \text{at } x + \theta a$$

by the mean value theorem, and this is, of course, $\geqslant 0$. Thus, $f(x) = (f(x) - f(x+a)) + f(x+a)$ is a decomposition of $f(x)$, so for $f(x)$ to be an extreme point we would have to have $f(x+a) = C_a f(x)$. Then $f'(x+a) = C_a f'(x)$, so $f'(x+a)/f(x+a) = f'(x)/f(x)$, i.e., $f'(x)/f(x)$ is a constant.

Thus, we have proved that the only possible extreme points are $\alpha e^{-\beta x}$. But we have not proved that they all are (or even that any of them are!) extreme. Even here, however, our general theorem scores for us. There has to be at least *one* extreme point, and since the map $f(x) \to af(bx)$, $a, b > 0$ takes extreme points into extreme points, the whole positive result follows. (Nice!)

With a little more work, using the fact that the extreme point always span the whole set, we could even go on and prove the representation theorem that every c.m. function $f(x)$ can be written as $\int_0^\infty e^{-tx} d\mu(t)$, $d\mu$ a positive measure.

109. If we look at all functions $f(m, n)$ with this property then they form a locally compact convex cone under the topology of uniform pointwise convergence (because at each neighboring point the bound is at most 4 times this point's value.) We wish to show that there is only one such function (up to constant multiples), so it suffices to show that there is only *one* extreme point. But if T denotes translation by $(1,0)$ and S is translation by $(0,1)$, we have $f = \frac{1}{4}Tf + \frac{1}{4}T^{-1}f + \frac{1}{4}Sf + \frac{1}{4}S^{-1}f$, so for an extreme point we must have $Tf = \lambda f$, $Sf = \mu f$. But then $f = \frac{1}{4}(\lambda + 1/\lambda + \mu + 1/\mu)f$ so $\lambda + 1/\lambda + \mu + 1/\mu = 4$, so $\lambda = \mu = 1$, and hence f is a constant.

Printed in Dunstable, United Kingdom

82978438R00070